Natural Language Understanding with Python

Combine natural language technology, deep learning, and large language models to create human-like language comprehension in computer systems

Deborah A. Dahl

BIRMINGHAM—MUMBAI

Natural Language Understanding with Python

Group Product Manager: Ali Abidi

Senior Editor: Tazeen Shaikh

Technical Editor: Rahul Limbachiya

Copy Editor: Safis Editing

Project Coordinator: Farheen Fathima

Proofreader: Safis Editing

Indexer: Rekha Nair

Production Designer: Joshua Misquitta

Marketing Coordinators: Shifa Ansari and Vinishka Kalra

First published: June 2023

Production reference: 1230623

Published by Packt Publishing Ltd.

Livery Place

35 Livery Street

Birmingham

B3 2PB, UK.

ISBN 978-1-80461-342-9

www.packtpub.com

This book is dedicated to my grandchildren, Freddie and Matilda, two tireless explorers who never stop surprising me and whose endless curiosity is a continual inspiration.

Contributors

About the author

Deborah A. Dahl is the principal at Conversational Technologies, with over 30 years of experience in natural language understanding technology. She has developed numerous natural language processing systems for research, commercial, and government applications, including a system for NASA, and speech and natural language components on Android. She has taught over 20 workshops on natural language processing, consulted on many natural language processing applications for her customers, and written over 75 technical papers. This is Deborah's fourth book on natural language understanding topics. Deborah has a PhD in linguistics from the University of Minnesota and postdoctoral studies in cognitive science from the University of Pennsylvania.

About the reviewers

Krishnan Raghavan is an IT professional with over 20+ years of experience in the area of software development and delivery excellence across multiple domains and technology, ranging from C++ to Java, Python, data warehousing, and big data tools and technologies.

When not working, Krishnan likes to spend time with his wife and daughter besides reading fiction, nonfiction, and technical books. Krishnan tries to give back to the community by being part of GDG, a Pune volunteer group, helping the team to organize events. Currently, he is unsuccessfully trying to learn how to play the guitar.

You can connect with Krishnan at mailtokrishnan@gmail.com or via LinkedIn at www.linkedin.com/in/krishnan-raghavan.

I would like to thank my wife, Anita, and daughter, Ananya, for giving me the time and space to review this book.

Mannai Mortadha is a dedicated and ambitious individual known for his strong work ethic and passion for continuous learning. Born and raised in an intellectually stimulating environment, Mannai developed a keen interest in exploring the world through knowledge and innovation. Mannai's educational journey began with a focus on computer science and technology. He pursued a degree in computer engineering, where he gained a solid foundation in programming, artificial intelligence algorithms, and system design. With a natural aptitude for problem-solving and an innate curiosity for cutting-edge technologies, Mannai excelled in his studies and consistently sought opportunities to expand his skills. Throughout his academic years, Mannai actively engaged in various extracurricular activities, including participating in hackathons and coding competitions. These experiences in Google, Netflix, Microsoft, and Talan Tunisie not only honed his technical abilities but also fostered his collaborative and teamwork skills.

Acknowledgment

I have worked as an independent consultant for a large part of my career. One of the benefits of working independently has been that I've been able to work with colleagues in many different organizations, which has exposed me to a rich set of technical perspectives that I would not have gotten from working at a single company or even a few companies.

I would like to express my gratitude to the colleagues, too many to mention individually, that I have worked with throughout my career, from my student days forward. These include colleagues at the University of Illinois, the University of Minnesota, the University of Pennsylvania, Unisys Corporation, MossRehab, Psycholinguistic Technologies, Autism Language Therapies, Openstream, the World Wide Web Consortium, the World Wide Web Foundation, the Applied Voice Input-Output Society, Information Today, New Interactions, the University Space Research Association, NASA Ames Research Center, and the Open Voice Network. I've learned so much from you.

Table of Contents

Preface   xv

Part 1: Getting Started with Natural Language Understanding Technology

1

Natural Language Understanding, Related Technologies, and Natural Language Applications   3

Understanding the basics of natural language	4	Classification	11
		Sentiment analysis	11
Global considerations – languages, encodings, and translations	4	Spam and phishing detection	11
		Fake news detection	12
The relationship between conversational AI and NLP	6	Document retrieval	12
		Analytics	12
Exploring interactive applications – chatbots and voice assistants	7	Information extraction	13
		Translation	13
Generic voice assistants	8	Summarization, authorship, correcting grammar, and other applications	13
Enterprise assistants	8	A summary of the types of applications	14
Translation	9		
Education	9	A look ahead – Python for NLP	15
Exploring non-interactive applications	10	Summary	15

2

Identifying Practical Natural Language Understanding Problems   17

Identifying problems that are the appropriate level of difficulty for the technology	18	Taking maintenance costs into account	31
Looking at difficult applications of NLU	21	A flowchart for deciding on NLU applications	33
Looking at applications that don't need NLP	26	Summary	34
Training data	29		
Application data	31		

Taking development costs into account   31

Part 2: Developing and Testing Natural Language Understanding Systems

3

Approaches to Natural Language Understanding – Rule-Based Systems, Machine Learning, and Deep Learning   37

Rule-based approaches	38	Representing documents	43
Words and lexicons	38	Classification	44
Part-of-speech tagging	38	Deep learning approaches	44
Grammar	39	Pre-trained models	45
Parsing	39	Considerations for selecting technologies	46
Semantic analysis	40	Summary	47
Pragmatic analysis	41		
Pipelines	42		

Traditional machine learning approaches   42

4

Selecting Libraries and Tools for Natural Language Understanding   49

Technical requirements	50	Developing software – JupyterLab and GitHub	51
Installing Python	50		

JupyterLab   51
GitHub   53

Exploring the libraries   **53**
Using NLTK   53
Using spaCy   55
Using Keras   57
Learning about other NLP libraries   58
Choosing among NLP libraries   58

Learning about other packages useful for
NLP   59

Looking at an example   **60**
Setting up JupyterLab   60
Processing one sentence   62
Looking at corpus properties   63

Summary   **68**

5

Natural Language Data – Finding and Preparing Data   69

**Finding sources of data and
annotating it**   **69**
Finding data for your own application   70
Finding data for a research project   71
Metadata   73
Generally available corpora   74

**Ensuring privacy and observing
ethical considerations**   **75**
Ensuring the privacy of training data   76
Ensuring the privacy of runtime data   76
Treating human subjects ethically   76
Treating crowdworkers ethically   76

Preprocessing data   **77**
Removing non-text   77

Regularizing text   80
Spelling correction   88

**Application-specific types of
preprocessing**   **89**
Substituting class labels for words and
numbers   89
Redaction   90
Domain-specific stopwords   90
Remove HTML markup   90
Data imbalance   91
Using text preprocessing pipelines   91

**Choosing among preprocessing
techniques**   **91**
Summary   **93**

6

Exploring and Visualizing Data   95

Why visualize?   **96**
Text document dataset – Sentence Polarity
Dataset   96

Data exploration   **97**
Frequency distributions   97
Measuring the similarities among
documents   113

**General considerations for
developing visualizations**   **119**
**Using information from
visualization to make decisions
about processing**   **123**
Summary   **124**

7

Selecting Approaches and Representing Data   125

Selecting NLP approaches	126	Understanding vectors for document representation	131
Fitting the approach to the task	126		
Starting with the data	126	**Representing words with context-independent vectors**	137
Considering computational efficiency	127		
Initial studies	128	Word2Vec	137
Representing language for NLP applications	128	**Representing words with context-dependent vectors**	141
Symbolic representations	128	**Summary**	141
Representing language numerically with vectors	131		

8

Rule-Based Techniques   143

Rule-based techniques	143	Lemmatization	148
Why use rules?	144	Ontologies	149
Exploring regular expressions	145	**Sentence-level analysis**	151
Recognizing, parsing, and replacing strings with regular expressions	145	Syntactic analysis	152
General tips for using regular expressions	148	Semantic analysis and slot filling	155
Word-level analysis	148	**Summary**	161

9

Machine Learning Part 1 – Statistical Machine Learning   163

A quick overview of evaluation	164	**Classifying documents with Support Vector Machines (SVMs)**	169
Representing documents with TF-IDF and classifying with Naïve Bayes	165	**Slot-filling with CRFs**	171
Summary of TF-IDF	165	Representing slot-tagged data	172
Classifying texts with Naïve Bayes	165	**Summary**	177
TF-IDF/Bayes classification example	166		

10

Machine Learning Part 2 – Neural Networks and Deep Learning Techniques   179

Basics of NNs	180	Looking at another approach – CNNs	192
Example – MLP for classification	183		
Hyperparameters and tuning	190	Summary	193
Moving beyond MLPs – RNNs	191		

11

Machine Learning Part 3 – Transformers and Large Language Models   195

Technical requirements	196	Defining the model for fine-tuning	203
Overview of transformers and LLMs	196	Defining the loss function and metrics	204
		Defining the optimizer and the number of epochs	204
Introducing attention	197		
Applying attention in transformers	198	Compiling the model	205
Leveraging existing data – LLMs or pre-trained models	198	Training the model	206
		Plotting the training process	207
BERT and its variants	198	Evaluating the model on the test data	209
Using BERT – a classification example	199	Saving the model for inference	209
		Cloud-based LLMs	210
Installing the data	201	ChatGPT	211
Splitting the data into training, validation, and testing sets	202	Applying GPT-3	212
Loading the BERT model	202	Summary	214

12

Applying Unsupervised Learning Approaches   215

What is unsupervised learning?	216	techniques and label derivation	217
Topic modeling using clustering		Grouping semantically similar documents	217

Applying BERTopic to 20 newsgroups   219
After clustering and topic labeling   226

Making the most of data with weak
supervision   226
Summary   227

13

How Well Does It Work? – Evaluation   229

Why evaluate an NLU system?   229
Evaluation paradigms   231
Comparing system results on standard
metrics   232
Evaluating language output   232
Leaving out part of a system – ablation   232
Shared tasks   233

Data partitioning   233
Evaluation metrics   235
Accuracy and error rate   235
Precision, recall, and F_1 236
The receiver operating characteristic and
area under the curve   237

Confusion matrix   238
User testing   239
Statistical significance of
differences   240
Comparing three text classification
methods   241
A small transformer system   241
TF-IDF evaluation   246
A larger BERT model   248
Summary   250

Part 3: Systems in Action – Applying Natural Language Understanding at Scale

14

What to Do If the System Isn't Working   253

Technical requirements   254
Figuring out that a system isn't
working   254
Initial development   254
Fixing accuracy problems   263

Changing data   264
Restructuring an application   270
Moving on to deployment   273
Problems after deployment   273
Summary   274

15

Summary and Looking to the Future 275

Overview of the book   275

Potential for improvement – better
accuracy and faster training   277

Better accuracy   277
Faster training   278
Other areas for improvement   278

Applications that are beyond the
current state of the art   280

Processing very long documents   281
Understanding and creating videos   281
Interpreting and generating sign
languages   281

Writing compelling fiction   282

Future directions in NLU technology
and research   283

Quickly extending NLU technologies to new
languages   283
Real-time speech-to-speech translation   284
Multimodal interaction   285
Detecting and correcting bias   285

Summary   286
Further reading   286

Index   287

Other Books You May Enjoy   300

Preface

Natural language understanding (**NLU**) is a technology that structures language so that computer systems can further process it to perform useful applications.

Developers will find that this practical guide enables them to use NLU techniques to develop many kinds of NLU applications, and managers will be able to identify areas where NLU can be applied to solve real problems in their enterprises.

Complete with step-by-step explanations of essential concepts and practical examples, you will begin by learning what NLU is and how it can be applied. You will then learn about the wide range of current NLU techniques, and you will learn about the best situations to apply each one, including the new **large language models** (**LLMs**). In the process, you will be introduced to the most useful Python NLU libraries. Not only will you learn the basics of NLU, but you will also learn about many practical issues such as acquiring data, evaluating systems, improving your system's results, and deploying NLU applications. Most importantly, you will not just learn a rote list of techniques, but you will learn how to take advantage of the vast number of NLU resources on the web in your future work.

Who this book is for

Python developers who are interested in learning about NLU and applying **natural language processing** (**NLP**) technology to real problems will get the most from this book, including computational linguists, linguists, data scientists, NLP developers, conversational AI developers, and students of these topics. The earlier chapters will also be interesting for non-technical project managers.

Working knowledge of Python is required to get the best from this book. You do not need any previous knowledge of NLU.

What this book covers

This book includes fifteen chapters that will take you through a process that starts from understanding what NLU is, through selecting applications, developing systems, and figuring out how to improve a system you have developed.

Chapter 1, Natural Language Understanding, Related Technologies, and Natural Language Applications, provides an explanation of what NLU is, and how it differs from related technologies such as speech recognition.

Chapter 2, Identifying Practical Natural Language Understanding Problems, systematically goes through a wide range of potential applications of NLU and reviews the specific requirements of each type of application. It also reviews aspects of an application that might make it difficult for the current state of the art.

Chapter 3, Approaches to Natural Language Understanding – Rule-Based Systems, Machine Learning, and Deep Learning, provides an overview of the main approaches to NLU and discusses their benefits and drawbacks, including rule-based techniques, statistical techniques, and deep learning. It also discusses popular pre-trained models such as BERT and its variants. Finally, it discusses combining different approaches into a solution.

Chapter 4, Selecting Libraries and Tools for Natural Language Understanding, helps you get set up to process natural language. It begins by discussing general tools such as Jupyter Labs and GitHub, and how to install and use them. It then goes on to discuss installing Python and the many available Python libraries that are available for NLU. Libraries that are discussed include NLTK, spaCy, and TensorFlow/Keras.

Chapter 5, Natural Language Data – Finding and Preparing Data, teaches you how to identify and prepare data for processing with NLU techniques. It discusses data from databases, the web, and other documents as well as privacy and ethics considerations. The Wizard of Oz technique and other simulated data acquisition approaches, such as generation, are covered briefly. For those of you who don't have access to your own data, or to those who wish to compare their results to those of other researchers, this chapter also discusses generally available and frequently used corpora. It then goes on to discuss preprocessing steps such as stemming and lemmatization.

Chapter 6, Exploring and Visualizing Data, discusses exploratory techniques for getting an overall picture of the data such as summary statistics (word frequencies, category frequencies, and so on). It will also discuss visualization tools such as matplotlib. Finally, it discusses the kinds of decisions that can be made based on visualization and statistical results.

Chapter 7, Selecting Approaches and Representing Data, discusses considerations for selecting approaches, for example, amount of data, training resources, and intended application. This chapter also discusses representing language with such techniques as vectors and embeddings in preparation for quantitative processing. It also discusses combining multiple approaches through the use of pipelines.

Chapter 8, Rule-Based Techniques, discusses how to apply rule-based techniques to specific applications. Examples include regular expressions, lemmatization, syntactic parsing, semantic role assignment and ontologies. This chapter primarily uses the NLTK libraries.

Chapter 9, Machine Learning Part 1 - Statistical Machine Learning, discusses how to apply statistical machine techniques such as Naïve Bayes, TF-IDF, support vector machines and conditional random fields to tasks such as classification, intent recognition, and entity extraction. The emphasis will be on newer techniques such as SVM and how they provide improved performance over more traditional approaches.

Chapter 10, Machine Learning Part 2 – Neural Networks and Deep Learning Techniques, covers applying machine learning techniques based on neural networks (fully connected networks, CNNs and RNNs) to problems such classification and information extraction. The chapter compares results using these approaches to the approaches described in the previous chapter. The chapter discusses neural net concepts such as hyperparameters, learning rate, and training iterations. This chapter uses the TensorFlow/Keras libraries.

Chapter 11, Machine Learning Part 3 – Transformers and Large Language Models, covers the currently best-performing techniques in natural language processing – transformers and pretrained models. It discusses the insights behind transformers and include an example of using transformers for text classification. Code for this chapter is based on the TensorFlow/Keras Python libraries.

Chapter 12, Applying Unsupervised Learning Approaches, discusses applications of unsupervised learning, such as topic modeling, including the value of unsupervised learning for exploratory applications and maximizing scarce data. It also addresses types of partial supervision such as weak supervision and distant supervision.

Chapter 13, How Well Does It Work? – Evaluation, covers quantitative evaluation. This includes segmenting the data into training, validation and test data, evaluation with cross-validation, evaluation metrics such as precision and recall, area under the curve, ablation studies, statistical significance, and user testing.

Chapter 14, What to Do If the System Isn't Working, discusses system maintenance. If the original model isn't adequate or if the situation in the real world changes, how does the model have to be changed? The chapter discusses adding new data and changing the structure of the application while at the same time ensuring that new data doesn't degrade the performance of the existing system.

Chapter 15, Summary and Looking to the Future, provides an overview of the book and a look to the future. It discusses where there is potential for improvement in performance as well as faster training, more challenging applications, and future directions for technology as well as research in this exciting technology.

To get the most out of this book

The code for this book is provided in the form of Jupyter Notebooks. To run the notebooks, you should have a comfortable understanding of coding in Python and be familiar with some basic libraries. Additionally, you'll need to install the required packages.

The easiest way to install them is by using Pip, a great package manager for Python. If Pip is not yet installed on your system, you can find the installation instructions here: `https://pypi.org/project/pip/`.

Working knowledge of the Python programming language will assist with understanding the key concepts covered in this book. The examples in this book don't require GPUs and can run on CPUs, although some of the more complex machine learning examples would run faster on a computer with a GPU.

The code for this book has *only been tested on Windows 11* (64-bit).

Software/hardware used in the book	Operating system requirements
Basic platform tools	
Python 3.9	Windows, macOS, or Linux
Jupyter Notebooks	
pip	
Natural Language Processing and Machine Learning	
NLTK	Windows, macOS, or Linux
spaCy and displaCy	
Keras	
TensorFlow	
Scikit-learn	
Graphing and visualization	
Matplotlib	Windows, macOS, or Linux
Seaborn	

Download the example code files

You can download the example code files for this book from GitHub at `https://github.com/PacktPublishing/Natural-Language-Understanding-with-Python`. If there's an update to the code, it will be updated in the GitHub repository.

We also have other code bundles from our rich catalog of books and videos available at `https://github.com/PacktPublishing/`. Check them out!

Download the color images

We also provide a PDF file that has color images of the screenshots and diagrams used in this book. You can download it here: `https://packt.link/HrkNr`.

Conventions used

There are a number of text conventions used throughout this book.

`Code in text`: Indicates code words in text, database table names, folder names, filenames, file extensions, pathnames, dummy URLs, user input, and Twitter handles. Here is an example: "We'll model the adjacency matrix using the `ENCOAdjacencyDistributionModule` object."

A block of code is set as follows:

```
preds = causal_bert.inference(
    texts=df['text'],
    confounds=df['has_photo'],
)[0]
```

Any command-line input or output is written as follows:

```
$ pip install demoji
```

Bold: Indicates a new term, an important word, or words that you see onscreen. For instance, words in menus or dialog boxes appear in **bold**. Here is an example: "Select **System info** from the **Administration** panel."

> **Tips or important notes**
> Appear like this.

Get in touch

Feedback from our readers is always welcome.

General feedback: If you have questions about any aspect of this book, email us at customercare@packtpub.com and mention the book title in the subject of your message.

Errata: Although we have taken every care to ensure the accuracy of our content, mistakes do happen. If you have found a mistake in this book, we would be grateful if you would report this to us. Please visit www.packtpub.com/support/errata and fill in the form.

Piracy: If you come across any illegal copies of our works in any form on the internet, we would be grateful if you would provide us with the location address or website name. Please contact us at copyright@packt.com with a link to the material.

If you are interested in becoming an author: If there is a topic that you have expertise in and you are interested in either writing or contributing to a book, please visit authors.packtpub.com.

Share your thoughts

Once you've read *Natural Language Understanding with Python*, we'd love to hear your thoughts! Scan the QR code below to go straight to the Amazon review page for this book and share your feedback.

https://packt.link/r/1-804-61342-8

Your review is important to us and the tech community and will help us make sure we're delivering excellent quality content.

Download a free PDF copy of this book

Thanks for purchasing this book!

Do you like to read on the go but are unable to carry your print books everywhere?

Is your eBook purchase not compatible with the device of your choice?

Don't worry, now with every Packt book you get a DRM-free PDF version of that book at no cost.

Read anywhere, any place, on any device. Search, copy, and paste code from your favorite technical books directly into your application.

The perks don't stop there, you can get exclusive access to discounts, newsletters, and great free content in your inbox daily

Follow these simple steps to get the benefits:

1. Scan the QR code or visit the link below

https://packt.link/free-ebook/9781804613429

2. Submit your proof of purchase

3. That's it! We'll send your free PDF and other benefits to your email directly

Part 1:
Getting Started with Natural Language Understanding Technology

In *Part 1*, you will learn about natural language understanding and its applications. You will also learn how to decide whether natural language understanding is applicable to a particular problem. In addition, you will learn about the relative costs and benefits of different NLU techniques.

This part comprises the following chapters:

- *Chapter 1, Natural Language Understanding, Related Technologies, and Natural Language Applications*
- *Chapter 2, Identifying Practical Natural Language Understanding Problems*

1

Natural Language Understanding, Related Technologies, and Natural Language Applications

Natural language, in the form of both speech and writing, is how we communicate with other people. The ability to communicate with others using natural language is an important part of what makes us full members of our communities. The first words of young children are universally celebrated. Understanding natural language usually appears effortless, unless something goes wrong. When we have difficulty using language, either because of illness, injury, or just by being in a foreign country, it brings home how important language is in our lives.

In this chapter, we will describe natural language and the kinds of useful results that can be obtained from processing it. We will also situate **natural language processing** (**NLP**) within the ecosystem of related conversational AI technologies. We will discuss where natural language occurs (documents, speech, free text fields of databases, etc.), talk about specific natural languages (English, Chinese, Spanish, etc.), and describe the technology of NLP, introducing Python for NLP.

The following topics will be covered in this chapter:

- Understanding the basics of natural language
- Global considerations
- The relationship between conversational AI and NLP
- Exploring interactive applications
- Exploring non-interactive applications
- A look ahead – Python for NLP

Learning these topics will give you a general understanding of the field of NLP. You will learn what it can be used for, how it is related to other conversational AI topics, and the kinds of problems it can address. You will also learn about the many potential benefits of NLP applications for both end users and organizations.

After reading this chapter, you will be prepared to identify areas of NLP technology that are applicable to problems that you're interested in. Whether you are an entrepreneur, a developer for an organization, a student, or a researcher, you will be able to apply NLP to your specific needs.

Understanding the basics of natural language

We don't yet have any technologies that can extract the rich meaning that humans experience when they understand natural language; however, given specific goals and applications, we will find that the current state of the art can help us achieve many practical, useful, and socially beneficial results through NLP.

Both spoken and written languages are ubiquitous and abundant. *Spoken language* is found in ordinary conversations between people and intelligent systems, as well as in media such as broadcasts, films, and podcasts. *Written language* is found on the web, in books, and in communications between people such as emails. Written language is also found in the free text fields of forms and databases that may be available online but are not indexed by search engines (the **invisible web**).

All of these forms of language, when analyzed, can form the basis of countless types of applications. This book will lay the basis for the fundamental analysis techniques that will enable you to make use of natural language in many different applications.

Global considerations – languages, encodings, and translations

There are thousands of natural languages, both spoken and written, in the world, although the majority of people in the world speak one of the top 10 languages, according to Babbel.com (https://www.babbel.com/en/magazine/the-10-most-spoken-languages-in-the-world). In this book, we will focus on major world languages, but it is important to be aware that different languages can raise different challenges for NLP applications. For example, the written form of Chinese does not include spaces between words, which most NLP tools use to identify words in a text. This means that to process Chinese language, additional steps beyond recognizing whitespace are necessary to separate Chinese words. This can be seen in the following example, translated by Google Translate, where there are no spaces between the Chinese words:

与大多数西方语言不同，书面中文不使用空格分隔单词

Figure 1.1 – Written Chinese does not separate words with spaces, unlike most Western languages

Another consideration to keep in mind is that some languages have many different forms of the same word, with different endings that provide information about its specific properties, such as the role the word plays in a sentence. If you primarily speak English, you might be used to words with very few endings. This makes it relatively easy for applications to detect multiple occurrences of the same word. However, this does not apply to all languages.

For example, in English, the word *walked* can be used in different contexts with the same form but different meanings, such as *I walked*, *they walked*, or *she has walked*, while in Spanish, the same verb (*caminar*) would have different forms, such as *Yo caminé, ellos caminaron*, or *ella ha caminado*. The consequence of this for NLP is that additional preprocessing steps might be required to successfully analyze text in these languages. We will discuss how to add these preprocessing steps for languages that require them in *Chapter 5*.

Another thing to keep in mind is that the availability and quality of processing tools can vary greatly across languages. There are generally reasonably good tools available for major world languages such as Western European and East Asian languages. However, languages with fewer than 10 million speakers or so may not have any tools, or the available tools might not be very good. This is due to factors such as the availability of training data as well as reduced commercial interest in processing these languages.

Languages with relatively few development resources are referred to as **low-resourced languages**. For these languages, there are not enough examples of the written language available to train large machine learning models in standard ways. There may also be very few speakers who can provide insights into how the language works. Perhaps the languages are endangered, or they are simply spoken by a small population. Techniques to develop natural language technology for these languages are actively being researched, although it may not be possible or may be prohibitively expensive to develop natural language technology for some of these languages.

Finally, many widely spoken languages do not use Roman characters, such as Chinese, Russian, Arabic, Thai, Greek, and Hindi, among many others. In dealing with languages that use non-Roman alphabets, it's important to recognize that tools have to be able to accept different character encodings. **Character encodings** are used to represent the characters in different writing systems. In many cases, the functions in text processing libraries have parameters that allow developers to specify the appropriate encoding for the texts they intend to process. In selecting tools for use with languages that use non-Roman alphabets, the ability to handle the required encodings must be taken into account.

The relationship between conversational AI and NLP

Conversational artificial intelligence is the broad label for an ecosystem of cooperating technologies that enable systems to conduct spoken and text-based conversations with people. These technologies include speech recognition, NLP, dialog management, natural language generation, and text-to-speech generation. It is important to distinguish these technologies, since they are frequently confused. While this book will focus on NLP, we will briefly define the other related technologies so that we can see how they all fit together:

- **Speech recognition**: This is also referred to as **speech-to-text** or **automatic speech recognition** (**ASR**). Speech recognition is the technology that starts with spoken audio and converts it to text.

- **NLP**: This starts with written language and produces a structured representation that can be processed by a computer. The input written language can either be the result of speech recognition or text that was originally produced in written form. The structured format can be said to express a user's **intent** or purpose.

- **Dialog management**: This starts with the structured output of NLP and determines how a system should react. System reactions can include such actions as providing information, playing media, or getting more information from a user in order to address their intent.

- **Natural language generation**: This is the process of creating textual information that expresses the dialog manager's feedback to a user in response to their utterance.

- **Text-to-speech**: Based on the textural input created by the natural language generation process, the text-to-speech component generates spoken audio output when given text.

The relationships among these components are shown in the following diagram of a complete spoken dialog system. This book focuses on the NLP component. However, because many natural language applications use other components, such as speech recognition, text-to-speech, natural language generation, and dialog management, we will occasionally refer to them:

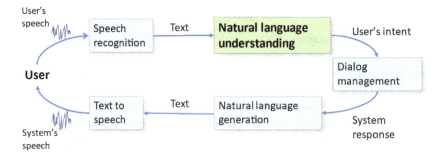

Figure 1.2 – A complete spoken dialog system

In the next two sections, we'll summarize some important natural language applications. This will give you a taste of the potential of the technologies that will be covered in this book, and it will hopefully get you excited about the results that you can achieve with widely available tools.

Exploring interactive applications – chatbots and voice assistants

We can broadly categorize NLP applications into two categories, namely **interactive applications**, where the fundamental unit of analysis is most typically a conversation, and **non-interactive applications**, where the unit of analysis is a document or set of documents.

Interactive applications include those where a user and a system are talking or texting to each other in real time. Familiar interactive applications include chatbots and voice assistants, such as smart speakers and customer service applications. Because of their interactive nature, these applications require very fast, almost immediate, responses from a system because the user is present and waiting for a response. Users will typically not tolerate more than a couple of seconds' delay, since this is what they're used to when talking with other people. Another characteristic of these applications is that the user inputs are normally quite short, only a few words or a few seconds long in the case of spoken interaction. This means that analysis techniques that depend on having a large amount of text available will not work well for these applications.

An implementation of an interactive application will most likely need one or more of the other components from the preceding system diagram, in addition to NLP itself. Clearly, applications with spoken input will need speech recognition, and applications that respond to users with speech or text will require natural language generation and text-to-speech (if the system's responses are spoken). Any application that does more than answer single questions will need some form of dialog management as well so that it can keep track of what the user has said in previous utterances, taking that information into account when interpreting later utterances.

Intent recognition is an important aspect of interactive natural language applications, which we will be discussing in detail in *Chapter 9* and *Chapter 14*. An intent is essentially a user's goal or purpose in making an utterance. Clearly, knowing what the user intended is central to providing the user with correct information. In addition to the intent, interactive applications normally have a requirement to also identify **entities** in user inputs, where entities are pieces of additional information that the system needs in order to address the user's intent. For example, if a user says, *"I want to book a flight from Boston to Philadelphia,"* the intent would be *make a flight reservation*, and the relevant entities are the departure and destination cities. Since the travel dates are also required in order to book a flight, these are also entities. Because the user didn't mention the travel dates in this utterance, the system should then ask the user about the dates, in a process called **slot filling**, which will be discussed in *Chapter 8*. The relationships between entities, intents, and utterances can be seen graphically in *Figure 1.3*:

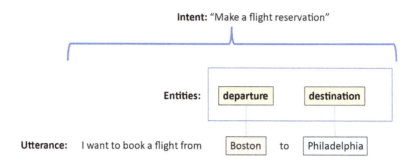

Figure 1.3 – The intent and entities for a travel planning utterance

Note that the intent applies to the overall meaning of the utterance, but the entities represent the meanings of only specific pieces of the utterance. This distinction is important because it affects the choice of machine learning techniques used to process these kinds of utterances. *Chapter 9*, will go into this topic in more detail.

Generic voice assistants

The **generic voice assistants** that are accessed through smart speakers or mobile phones, such as Amazon Alexa, Apple Siri, and Google Assistant, are familiar to most people. Generic assistants are able to provide users with general information, including sports scores, news, weather, and information about prominent public figures. They can also play music and control the home environment. Corresponding to these functions, the kinds of intents that generic assistants recognize are intents such as *get weather forecast for <location>*, where *<location>* represents an entity that helps fill out the *get weather forecast* intent. Similarly, *"What was the score for <team name> game?"* has the intent *get game score*, with the particular team's name as the entity. These applications have broad but generally shallow knowledge. For the most part, their interactions with users are just based on one or, at most, a couple of related inputs – that is, for the most part, they aren't capable of carrying on an extended conversation.

Generic voice assistants are mainly closed and proprietary. This means that there is very little scope for developers to add general capabilities to the assistant, such as adding a new language. However, in addition to the aforementioned proprietary assistants, an open source assistant called **Mycroft** is also available, which allows developers to add capabilities to the underlying system, not just use the tools that the platforms provide.

Enterprise assistants

In contrast to the generic voice assistants, some interactive applications have deep information about a specific company or other organization. These are **enterprise assistants**. They're designed to perform tasks specific to a company, such as customer service, or to provide information about a government or educational organization. They can do things such as check the status of an order, give bank customers account information, or let utility customers find out about outages. They are often connected to extensive databases of customer or product information; consequently, based on this information, they can provide deep but mainly narrow information about their areas of expertise. For example, they can tell you whether a particular company's products are in stock, but they don't know the outcome of your favorite sports team's latest game, which generic assistants are very good at.

Enterprise voice assistants are typically developed with toolkits such as the Alexa Skills Kit, Microsoft LUIS, Google Dialogflow, or Nuance Mix, although there are open source toolkits such as RASA (`https://rasa.com/`). These toolkits are very powerful and easy to use. They only require developers to give toolkits examples of the intents and entities that the application will need to find in users' utterances in order to understand what they want to do.

Similarly, text-based chatbots can perform the same kinds of tasks that voice assistants perform, but they get their information from users in the form of text rather than voice. Chatbots are becoming increasingly common on websites. They can supply much of the information available on the website, but because the user can simply state what they're interested in, they save the user from having to search through a possibly very complex website. The same toolkits that are used for voice assistants can also be used in many cases to develop text-based chatbots.

In this book, we will not spend too much time on the commercial toolkits because there is very little coding needed to create usable applications. Instead, we'll focus on the technologies that underly the commercial toolkits, which will enable developers to implement applications without relying on commercial systems.

Translation

The third major category of an interactive application is **translation**. Unlike the assistants described in the previous sections, translation applications are used to assist users to communicate with other people – that is, the user isn't having a conversation with the assistant but with another person. In effect, the applications perform the role of an interpreter. The application translates between two different human languages in order to enable two people who don't speak a common language to talk with each other. These applications can be based on either spoken or typed input. Although spoken input is faster and more natural, if speech recognition errors (which are common) occur, this can significantly interfere with the smoothness of communication between people.

Interactive translation applications are most practical when the conversation is about simple topics such as tourist information. More complex topics – for example, business negotiations – are less likely to be successful because their complexity leads to more speech recognition and translation errors.

Education

Finally, **education** is an important application of interactive NLP. Language learning is probably the most natural educational application. For example, there are applications that help students converse in a new language that they're learning. These applications have advantages over the alternative of practicing conversations with other people because applications don't get bored, they're consistent, and users won't be as embarrassed if they make mistakes. Other educational applications include assisting students with learning to read, learning grammar, or tutoring in any subject.

Figure 1.4 is a graphical summary of the different kinds of interactive applications and their relationships:

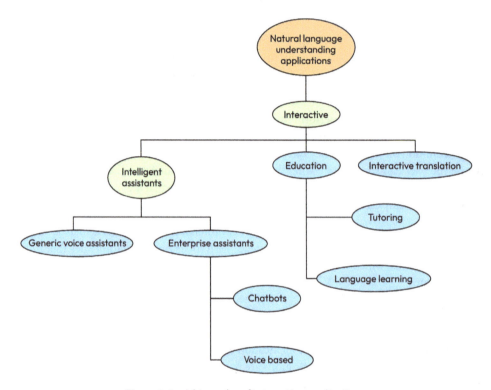

Figure 1.4 – A hierarchy of interactive applications

So far, we've covered interaction applications, where an end user is directly speaking to an NLP system, or typing into it, in real time. These applications are characterized by short user inputs that need quick responses. Now, we will turn to non-interactive applications, where speech or text is analyzed when there is no user present. The material to be analyzed can be arbitrarily long, but the processing time does not have to be immediate.

Exploring non-interactive applications

The other major type of natural language application is non-interactive, or offline applications. The primary work done in these applications is done by an NLP component. The other components in the preceding system diagram are not normally needed. These applications are performed on existing text, without a user being present. This means that real-time processing is not necessary because the user isn't waiting for an answer. Similarly, the system doesn't have to wait for the user to decide what to say so that, in many cases, processing can occur much more quickly than in the case of an interactive application.

Classification

A very important and widely used class of non-interactive natural language applications is document **classification**, or assigning documents to categories based on their content. Classification has been a major application area in NLP for many years and has been addressed with a wide variety of approaches.

One simple example of classification is a web application that answers customers' **frequently asked questions** (**FAQs**) by classifying a query into one of a set of given categories and then providing answers that have been previously prepared for each category. For this application, a classification system would be a better solution than simply allowing customers to select their questions from a list because an application could sort questions into hundreds of FAQ categories automatically, saving the customer from having to scroll through a huge list of categories. Another example of an interesting classification problem is automatically assigning genres to movies – for example, based on reviews or plot summaries.

Sentiment analysis

Sentiment analysis is a specialized type of classification where the goal is to classify texts such as product reviews into those that express positive and negative sentiments. It might seem that just looking for positive or negative words would work for sentiment analysis, but in this example, we can see that despite many negative words and phrases (`concern`, `break`, `problem`, `issues`, `send back`, and `hurt my back`), the review is actually positive:

"I was concerned that this chair, although comfortable, might break before I had it for very long because the legs were so thin. This didn't turn out to be a problem. I thought I might have to send it back. I haven't had any issues, and it's the one chair I have that doesn't hurt my back."

More sophisticated NLP techniques, taking context into account, are needed to recognize that this is a positive review. Sentiment analysis is a very valuable application because it is difficult for companies to do this manually if there are thousands of existing product reviews and new product reviews are constantly being added. Not only do companies want to see how their products are viewed by customers, but it is also very valuable for them to know how reviews of competing products compare to reviews of their own products. If there are dozens of similar products, this greatly increases the number of reviews relevant to the classification. A text classification application can automate a lot of this process. This is a very active area of investigation in the academic NLP community.

Spam and phishing detection

Spam detection is another very useful classification application, where the goal is to sort email messages into messages that the user wants to see and spam that should be discarded. This application is not only useful but also challenging because spammers are constantly trying to circumvent spam detection algorithms. This means that spam detection techniques have to evolve along with new ways of creating spam. For example, spammers often misspell keywords that might normally indicate spam by substituting the numeral *1* for the letter *l*, or substituting the numeral *0* for the letter *o*. While humans have no trouble reading words that are misspelled in this way, keywords that the computer is looking for will no longer match, so spam detection techniques must be developed to find these tricks.

Closely related to spam detection is detecting messages attempting to phish a user or get them to click on a link or open a document that will cause malware to be loaded onto their system. Spam is, in most cases, just an annoyance, but phishing is more serious, since there can be extremely destructive consequences if the user clicks on a phishing link. Any techniques that improve the detection of phishing messages will, therefore, be very beneficial.

Fake news detection

Another very important classification application is **fake news detection**. Fake news refers to documents that look very much like real news but contain information that isn't factual and is intended to mislead readers. Like spam detection and phishing detection, fake news detection is challenging because people who generate fake news are actively trying to avoid detection. Detecting fake news is not only important for safeguarding reasons but also from a platform perspective, as users will begin to distrust platforms that consistently report fake news.

Document retrieval

Document retrieval is the task of finding documents that address a user's search query. The best example of this is a routine web search of the kind most of us do many times a day. Web searches are the most well-known example of document retrieval, but document retrieval techniques are also used in finding information in any set of documents – for example, in the free-text fields of databases or forms.

Document retrieval is based on finding good matches between users' queries and the stored documents, so analyzing both users' queries and documents is required. Document retrieval can be implemented as a keyword search, but simple keyword searches are vulnerable to two kinds of errors. First, keywords in a query might be intended in a different sense than the matching keywords in documents. For example, if a user is looking for a new pair of glasses, thinking of *eyeglasses*, they don't want to see results for drinking glasses. The other type of error is where relevant results are not found because keywords don't match. This might happen if a user uses just the keyword *glasses*, and results that might have been found with the keywords *spectacles* or *eyewear* might be missed, even if the user is interested in those. Using NLP technology instead of simple keywords can help provide more precise results.

Analytics

Another important and broad area of natural language applications is analytics. **Analytics** is an umbrella term for NLP applications that attempt to gain insights from text, often the transcribed text from spoken interactions. A good example is looking at the transcriptions of interactions between customers and call center agents to find cases where the agent was confused by the customer's question or provided wrong information. The results of analytics can be used in the training of call center agents. Analytics can also be used to examine social media posts to find trending topics.

Information extraction

Information extraction is a type of application where structured information, such as the kind of information that could be used to populate a database, is derived from text such as newspaper articles. Important information about an event, such as the date, time, participants, and locations, can be extracted from texts reporting news. This information is quite similar to the intents and entities discussed previously when we talked about chatbots and voice assistants, and we will find that many of the same processing techniques are relevant to both types of applications.

An extra problem that occurs in information extraction applications is **named entity recognition** (**NER**), where references to real people, organizations, and locations are recognized. In extended texts such as newspaper articles, there are often multiple ways of referring to the same individual. For example, Joe Biden might be referred to as *the president, Mr. Biden, he*, or even *the former vice-president*. In identifying references to Joe Biden, an information extraction application would also have to avoid misinterpreting a reference to *Dr. Biden* as a reference to Joe Biden, since that would be a reference to his wife.

Translation

Translation between languages, also known as **machine translation**, has been one of the most important NLP applications since the field began. Machine translation hasn't been solved in general, but it has made enormous progress in the past few years. Familiar web applications such as Google Translate and Bing Translate usually do a very good job on text such as web pages, although there is definitely room for improvement.

Machine translation applications such as Google and Bing are less effective on other types of text, such as technical text that contains a great deal of specialized vocabulary or colloquial text of the kind that might be used between friends. According to Wikipedia (`https://en.wikipedia.org/wiki/Google_Translate`), Google Translate can translate 109 languages. However, it should be kept in mind that the accuracy for the less widely spoken languages is lower than that for the more commonly spoken languages, as discussed in the *Global considerations* section.

Summarization, authorship, correcting grammar, and other applications

Just as there are many reasons for humans to read and understand texts, there are also many applications where systems that are able to read and understand text can be helpful. Detecting plagiarism, correcting grammar, scoring student essays, and determining the authorship of texts are just a few. Summarizing long texts is also very useful, as is simplifying complex texts. Summarizing and simplifying text can also be applied when the original input is non-interactive speech, such as podcasts, YouTube videos, or broadcasts.

Figure 1.5 is a graphical summary of the discussion of non-interactive applications:

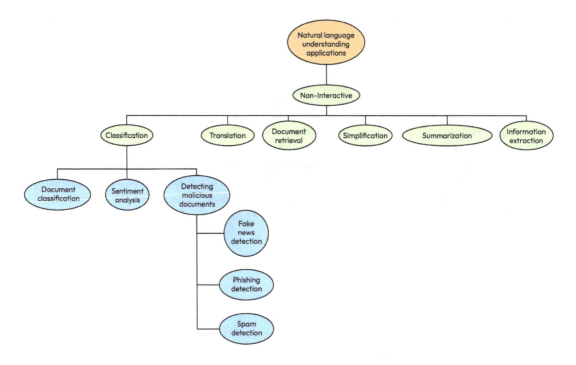

Figure 1.5 – A hierarchy of non-interactive applications

Figure 1.5 shows how the non-interactive NLP applications we've been discussing are related to each other. It's clear that classification is a major application area, and we will look at it in depth in *Chapter 9*, *Chapter 10*, and *Chapter 11*.

A summary of the types of applications

In the previous sections, we saw how the different types of interactive and non-interactive applications we have discussed relate to each other. It is apparent that NLP can be applied to solving many different and important problems. In the rest of the book, we'll dive into the specific techniques that are appropriate for solving different kinds of problems, and you'll learn how to select the most effective technologies for each problem.

A look ahead – Python for NLP

Traditionally, NLP has been accomplished with a variety of computer languages, from early, special-purpose languages, such as Lisp and Prolog, to more modern languages, such as Java and now Python. Currently, Python is probably the most popular language for NLP, in part because interesting applications can be implemented relatively quickly and developers can rapidly get feedback on the results of their ideas.

Another major advantage of Python is the very large number of useful, well-tested, and well-documented Python libraries that can be applied to NLP problems. Some of these libraries are NLTK, spaCy, scikit-learn, and Keras, to name only a few. We will be exploring these libraries in detail in the chapters to come. In addition to these libraries, we will also be working with development tools such as JupyterLab. You will also find other resources such as Stack Overflow and GitHub to be extremely valuable.

Summary

In this chapter, we learned about the basics of natural language and global considerations. We also looked at the relationship between conversational AI and NLP and explored interactive and non-interactive applications.

In the next chapter, we will be covering considerations concerning selecting applications of NLP. Although there are many ways that this technology can be applied, some possible applications are too difficult for the state of the art. Other applications that seem like good applications for NLP can actually be solved by simpler technologies. In the next chapter, you will learn how to identify these.

2

Identifying Practical Natural Language Understanding Problems

In this chapter, you will learn how to identify **natural language understanding** (NLU) problems that are a good fit for today's technology. That means they will not be too difficult for the state-of-the-art NLU approaches but neither can they be addressed by simple, non-NLU approaches. Practical NLU problems also require sufficient training data. Without sufficient training data, the resulting NLU system will perform poorly. The benefits of an NLU system also must justify its development and maintenance costs. While many of these considerations are things that project managers should think about, they also apply to students who are looking for class projects or thesis topics.

Before starting a project that involves NLU, the first question to ask is whether the goals of the project are a good fit for the current state of the art in NLU. Is NLU the right technology for solving the problem that you wish to address? How does the difficulty of the problem compare to the NLU state of the art?

Starting out, it's also important to decide what *solving the problem* means. Problems can be solved to different degrees. If the application is a class project, demo, or proof of concept, the solution does not have to be as accurate as a deployed solution that's designed for the robust processing of thousands of user inputs a day. Similarly, if the problem is a cutting-edge research question, any improvement over the current state of the art is valuable, even if the problem isn't completely solved by the work done in the project. How complete the solution has to be is a question that everyone needs to decide as they think about the problem that they want to address.

The project manager, or whoever is responsible for making the technical decisions about what technologies to use, should decide what level of accuracy they would find acceptable when the project is completed, keeping in mind that 100% accuracy is unlikely to be achievable in any natural language technology application.

This chapter will get into the details of identifying problems where NLU is applicable. Follow the principles discussed in this chapter, and you will be rewarded with a quality, working system that solves a real problem for its users.

The following topics are covered in this chapter:

- Identifying problems that are the appropriate level of difficulty for the technology
- Looking at difficult NLU applications
- Looking at applications that don't need NLP
- Training data
- Application data
- Taking development costs into account
- Taking maintenance costs into account
- A flowchart for deciding on NLU applications

Identifying problems that are the appropriate level of difficulty for the technology

> **Note**
>
> This chapter is focused on technical considerations. Questions such as whether a market exists for a proposed application, or how to decide whether customers will find it appealing, are important questions, but they are outside of the scope of this book.

Here are some kinds of problems that are a good fit for the state of the art.

Today's NLU is very good at handling problems based on specific, concrete topics, such as these examples:

- **Classifying customers' product reviews into positive and negative reviews**: Online sellers typically offer buyers a chance to review products they have bought, which is helpful for other prospective buyers as well as for sellers. But large online retailers with thousands of products are then faced with the problem of what to do with the information from thousands of reviews. It's impossible for human tabulators to read all the incoming reviews, so an automated product review classification system would be very helpful.

- **Answering basic banking questions about account balances or recent transactions**: Banks and other financial institutions have large contact centers that handle customer questions. Often, the most common reasons for calling are simple questions about account balances, which can be answered with a database lookup based on account numbers and account types. An automated system can handle these by asking callers for their account numbers and the kind of information they need.

- **Making simple stock trades**: Buying and selling stock can become very complex, but in many cases, users simply want to buy or sell a certain number of shares of a specific company. This kind of transaction only needs a few pieces of information, such as an account number, the company, the number of shares, and whether to buy or sell.

- **Package tracking**: Package tracking needs only a tracking number to tell users the status of their shipments. While web-based package tracking is common, sometimes, people don't have access to the web. With a natural language application, users can track packages with just a phone call.

- **Routing customers' questions to the right customer service agent**: Many customers have questions that can only be answered by a human customer service agent. For those customers, an NLU system can still be helpful by directing the callers to the call center agents in the right department. It can ask the customer the reason for their call, classify the request, and then automatically route their call to the expert or department that handles that topic.

- **Providing information about weather forecasts, sports scores, and historical facts**: These kinds of applications are characterized by requests that have a few well-defined parameters. For sports scores, this would be a team name and possibly the date of a game. For weather forecasts, the parameters include the location and timeframe for the forecast.

All of these applications are characterized by having unambiguous, correct answers. In addition, the user's language that the system is expected to understand is not too complex. These would all be suitable topics for an NLU project.

Let's illustrate what makes these applications suitable for today's technology by going into more detail on providing information about weather forecasts, sports scores, and historical facts.

Figure 2.1 shows a sample architecture for an application that can provide weather forecasts for different cities. Processing starts when the user asks, *What is the weather forecast for tomorrow in New York City?* Note that the user is making a single, short request, for specific information – the weather forecast, for a particular date, in a particular location. The NLU system needs to detect the intent (weather forecast), the entities' *location*, and the *date*. These should all be easy to find – the entities are very dissimilar, and the *weather forecast* intent is not likely to be confused with any other intents. This makes it straightforward for the NLU system to convert the user's question to a structured message that could be interpreted by a weather information web service, as shown at the top of the following figure:

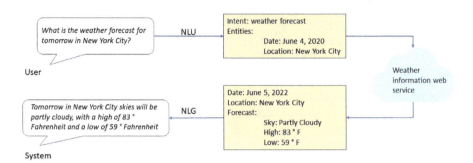

Figure 2.1 – A practical NLU application

Despite the fact that the information being requested is not very complex, there are many ways to ask about it, which means that it's not very practical to just make a list of possible user queries. *Table 2.1* illustrates a few of the many ways to make this request:

Some paraphrases of "What is the weather forecast for tomorrow in New York City?"
What will the weather be like tomorrow in New York?
What's tomorrow's weather for New York?
I want the New York City weather forecast for tomorrow.
The weather tomorrow in New York, please.
New York weather forecast for tomorrow.
Tomorrow's weather forecast for New York City.

Table 2.1 – Paraphrases for a weather request

Another aspect of this application that makes it a good candidate for NLU is that the information the user is asking about (weather forecasts) is available from multiple easily-accessible, cloud-based web services, with **application programming interfaces** (**APIs**) that are usually well documented. This makes it easy for developers to send queries to the web services and get back the information that the user requested in a structured form. This information can then be presented to the user. Developers have choices about how they want to present the information – for example, text, graphics, or a combination of text and graphics.

In *Figure 2.1*, we can see that the developer has chosen to present the information via natural language, and consequently, a **natural language generation** (**NLG**) component is used to generate the natural language output from a form. Other presentation options would be to show graphics, such as a picture of the sun partially covered by a cloud, or to simply show a form with the information received from the weather information web service. However, only the NLG option is a good fit for a spoken or voice-only interface such as a smart speaker since, with a voice-only interface, there is no way to display graphics.

The biggest benefit of NLU for an application such as weather forecasting is that NLU can handle the many possible ways that the user might ask this question with the same intent, as shown in *Table 2.1*.

Table 2.1 shows some paraphrases of a weather forecast request. These are just a few examples of possible ways to ask for a weather forecast. It is often surprising how many different ways there are to make even a simple request. If we could make a list of all the options, even if it was a very long list, NLU wouldn't be necessary.

We could theoretically just list all the possibilities and map them to the structured queries. However, it's actually very difficult to anticipate all the possible ways that someone would ask even a simple question about the weather. If a user happens to phrase their query in a way that the developer hasn't included in their list, the system will fail to respond. This can be very confusing to users because users won't understand why this query failed when similar queries worked. An NLU system will be able to cope with many more query variations.

As we've seen in this section, applications that have clear and easily identifiable intents and entities and that have definite answers that can be obtained from web resources, have a good chance of success with today's NLU technology.

Now, let's turn to applications that are unlikely to be successful because they require capabilities that are beyond the state of the art.

Looking at difficult applications of NLU

How can we tell whether the problem is too hard for the state of the art? First of all, we can ask what it means for a problem to be *too hard*. Here are some consequences of trying to use NLU for an application that is beyond the state of the art:

- The system will be unable to reliably understand user queries
- Answers will contain errors because the system has misunderstood user queries
- The system will have to say *I don't know* or *I can't do that* so frequently that users become frustrated and decide not to use the application anymore

It's important to keep in mind that the state of the art is rapidly improving. Remarkable progress has been made recently as cloud-based **large language models** (**LLMs**) such as ChatGPT have become available. Some applications that might be very hard now will not always be too hard.

Let's look at a few of the characteristics of today's difficult NLU problems.

Applications that require the system to use judgment or common sense

Unlike the weather example in the previous section, applications that require judgment are applications where there isn't a single correct answer, or even a few reasonable alternatives. These could include applications where the user is asking for advice that depends on many, often complex, considerations. Here are some examples:

- Should I learn Python?
- Should I get a COVID vaccine?
- Should I buy an electric car?
- Is this a good time to buy a house?

To answer the first question, the system needs specific knowledge about the user (whether the user already has a programming background or what they want to do with their new programming skills). LLM-based systems, such as ChatGPT, will respond to these kinds of questions in a general way – for example, by providing generic considerations about buying a house – but they can't give advice that's specific to the user, because they don't know anything about the user.

Applications in which the system is asked for a subjective opinion are also very difficult to handle well, such as these examples:

- What is the best movie of all time?
- Who was the most talented 20th-century actor?
- What is a good way to cook chicken that doesn't take more than half an hour?

To fully answer these kinds of queries requires the system to have a lot of general knowledge, such as actors who had careers in the 20th century. A system could respond to subjective questions by giving a random answer – just pick a movie at random and say that that movie is the best of all time. However, a randomly picked movie is not necessarily going to even be good, let alone the best movie of all time.

In that case, if there's a follow-up question, the system won't be able to explain or defend its opinion. So, if you asked a system *Should I buy an electric car*, and it said *Yes*, it wouldn't be able to explain why it said yes. In fact, it's probably too difficult for many of today's systems to even realize that they're being asked a subjective question. As in the case of questions that require the knowledge of the user to give a good answer, LLM-based systems will give generic answers to subjective questions, but they will admit that they aren't able to deal with subjectivity.

Applications that require dealing with hypotheticals, possibilities, and counterfactuals

Another difficult area is dealing with information that isn't true or is possibly not true. When the user asks about something that might happen, if the circumstances are right, the user is asking about a hypothetical or a possibility. Today's state-of-the-art systems are good at providing specific, concrete information, but the technology is not good at reasoning about possibilities. Here are some examples:

- If I have a budget of $15,000, how big of a patio should I be able to build, assuming I'm willing to do some of the work myself?
- If I have six people, how many pizzas should I get?
- If there's no rain in the forecast tomorrow, remind me to water my plants.

Similarly, systems aren't very good at reasoning about things that aren't true. For example, consider the sentence, *I'd like to find a nearby Asian restaurant, but not Japanese*. To answer this question correctly, the system has to find Asian restaurants, and it has to understand that it should exclude Japanese restaurants, which are nevertheless Asian, from the list.

Applications that require combining information from a language with information from various sensors

Some very interesting applications could involve integrating information from language and cameras or microphones. These are called **multimodal** applications because they integrate multiple modalities such as speech, images, and non-speech audio such as music:

- Is this cake done? (holding camera up to cake)
- What is this noise that my car is making? (holding microphone up to engine)

These applications are currently beyond the state of the art of today's commercial natural language technology, although they could be appropriate for an exploratory research project. They are also currently outside of the capabilities of LLMs, which can only understand text input.

Applications that integrate broad general or expert knowledge

When users interact with an NLU system, they have goals that they want to accomplish. In many cases, the system has some kind of knowledge or expertise that the user doesn't have, and the user wants to take advantage of that expertise. But where does that expertise come from? Providing systems with large amounts of knowledge is difficult. There are existing web APIs for simple information such as sports scores and weather. Systems such as Wolfram Alpha can also answer more complicated questions, such as scientific facts.

On the other hand, answering questions that require the use of expert knowledge, such as medical information, is more difficult, as there's no easily accessible source of this kind of knowledge. In addition, existing sources of information might be inconsistent. One obvious source of large amounts of knowledge is the **World Wide Web** (**WWW**), which is the major source of knowledge of LLM. However, knowledge available on the WWW can be wrong, inconsistent, or not applicable to a particular situation, so it has to be used with caution.

These are a few examples of difficult topics for today's natural language technology:

- **Answer complex technical questions**: A statement like *I can't connect to the internet* requires the system to have a lot of information about internet connectivity as well as how to debug connectivity problems. It would also have to have access to other time-sensitive information such as whether there are global internet outages in the user's area.

- **Answer questions that require an understanding of human relationships**: *My friend won't talk to me since I started dating her boyfriend; what should I do?* A system would have to understand a lot about dating, and probably dating in a specific culture as well, in order to give a good answer to a question like this.

- **Read a book and tell me whether I would like that book**: Today's systems would have a hard time even reading and understanding an entire book since long texts like books contain very complex information. In addition to just reading a book, for a system to tell me whether I would like it requires a lot of information about me and my reading interests.

- **Read an article from a medical journal and tell me whether the findings apply to me**: Answering questions like this would require a tremendous amount of information about the user's health and medical history, as well as the ability to understand medical language and interpret the results of medical studies.

- **Understand jokes**: Understanding jokes often requires considerable cultural knowledge. Think about what knowledge a system would need to be able to understand the traditional joke, *Why did the chicken cross the road? To get to the other side.* This is funny because the question leads the user to believe that the chicken has an interesting reason for crossing the road, but its reason turns out to be extremely obvious. Not only would it be very hard for a system to be able to understand why this particular joke is funny, but this is only one joke—just being able to understand this joke wouldn't help a system understand any other jokes.

- **Interpret figures of speech**: *I could eat a horse* doesn't mean that you want to eat a horse, it just means that you're very hungry. A system would have to realize that this is a figure of speech because horses are very large and no one could actually eat a horse in one sitting, no matter how hungry they are. On the other hand, *I could eat a pizza* is not a figure of speech and probably just means that the user would like to order a pizza.

- **Understand irony and sarcasm**: If a book review contains a sentence like *The author is a real genius*, the review writer might mean that the author is literally a genius, but not necessarily. This could be intended sarcastically to mean that the author is not a genius at all. If this sentence is followed by *My three-year-old could have written a better book*, we can tell that the first sentence was intended to be taken as sarcasm. NLU systems can't understand sarcasm. They also don't know that three-year-olds are unlikely to be able to write good books, and so the writer of the review is claiming that the book is worse than one authored by a three-year-old, and so it is a bad book.

- **Be able to make use of complex knowledge**: As an example of complex knowledge, consider the utterance, *My cake is as flat as a pancake; what went wrong?* To answer this question, the system has to understand that a cake shouldn't be flat but that pancakes are normally flat. It also has to understand that we're talking about a cake that has been baked, as unbaked cakes are typically flat. Once the system has figured all this out, it also has to understand the process of baking enough to give advice about why the cake is flat.

One general property shared by many of these difficult types of applications is that there often isn't any one data source where the answers can be obtained. That is, there aren't any backend data sources that developers can just query to answer a question like *Is this a good time to buy an electric car?* This is in contrast to the earlier weather forecast example, where developers can go to a single backend data source.

Rather than trying to find a single backend data source, one strategy might be to do a web search for the question. But as anyone who's done a web search knows, there will be millions of search results (nearly 2 billion for *Is this a good time to buy an electric car?*), and what's worse, the answers are not likely to be consistent with each other. Some pages will assert that it is a good time to buy an electric car, and others will assert that it is not. So, the strategy of using a web search to answer questions without a good data source will probably not work. However, being able to integrate information from across the web is a strength of LLMs, so if the information is available on the web, an LLM such as ChatGPT will be able to find it.

Applications where users often don't have a clear idea of what they want

Users don't always state their intentions very clearly. As an example, consider a tourist who's visiting an unfamiliar town. Perhaps the town provides a service that tourists can call to find out about public transportation options. If a tourist asks a question like *What train should I take to get from the Marriott Hotel to 123 Market Street?*, a literal answer might be *You can't take the train from the Marriott Hotel to 123 Market Street*. Or the user might be offered a circuitous route that takes six hours.

A human agent could figure out that the caller's actual goal is probably to get from the Marriott Hotel to 123 Market Street, and the reference to the train was just the caller's guess that the train would be a good way to do that. In that case, a human agent could say something like *There isn't really a good train route between those two locations; would you like some other ideas about how to get between them?* This would be natural for a human agent but very difficult for an automated system, because the system would need to be able to reason about what the user's real goal is.

Applications that require understanding multiple languages

As discussed in *Chapter 1*, language technology is better for some languages than others. If a system has to be able to communicate with users (by speech or text) in different languages, then language models for each language have to be developed. Processing for some languages will be more accurate than processing for other languages, and for some languages, processing might not be good enough at all. At the current state of the art, NLP technology for major European, Middle Eastern, and Asian languages should be able to handle most applications.

In some applications, the system has to be prepared to speak different languages depending on what the user says. To do this, the system has to be able to tell the different languages apart just by their sounds or words. This technology is called **language identification**. Identifying commonly spoken languages is not difficult but, again, this is not the case for less common languages.

In the case of languages with very little training data, such as languages with fewer than one million speakers, the language may not have been studied well enough for natural language applications to be developed for that language.

Even more difficult than understanding multiple languages is handling cases where two or more languages are mixed in the same sentence. This often happens when several different languages are spoken in the same area, and people can assume that anyone they talk with can understand all the local languages. Mixing languages in the same sentence is called **code-switching**. Processing sentences with code-switching is even more difficult than processing several languages in the same application because the system has to be prepared for any word in any of the languages it knows at any point in the sentence. This is a difficult problem for today's technology.

In the preceding discussion, we've reviewed many factors that make applications too difficult for today's state of the art in NLP. Let's now look at applications that are too easy.

Looking at applications that don't need NLP

Turning from applications that are too difficult, we can also look at applications that are too easy – that is, applications where simpler solutions than NLP will work, and where NLP is overkill. These are applications where the complexity of the problem doesn't justify the complexity of building and managing a natural language system.

Natural language is characterized by unpredictable inputs and an indirect mapping of words to meanings. Different words can have the same meaning, and different meanings can be expressed by the same words, depending on the context. If there is a simple one-to-one mapping between inputs and meanings, NLP isn't necessary.

Text that can be analyzed with regular expressions

The first case where NLU isn't necessary is when the possible inputs consist of a limited set of options, such as cities, states, or countries. Internally, such inputs can be represented as lists, and can be analyzed via table lookup. Even if there are synonyms for certain inputs (*UK* for the *United Kingdom*, for example), the synonyms can be included in the lists as well.

A slightly more complicated, but still simple, input is when every input to the system is composed according to easily stated, unvarying rules. NLP is not necessary in those cases because the input is predictable. Good examples of these kinds of simple expressions are telephone numbers, which have fixed, predictable formats, or dates, which are more varied, but still limited. In addition to these generic expressions, in specific applications, there is often a requirement to analyze expressions such as product IDs or serial numbers. These types of inputs can be analyzed with regular expressions. Regular expressions are rules that describe patterns of characters (alphabetical, numerical, or special characters). For example, the `^\d{5}(-\d{4})?$` regular expression matches US zip codes, either containing five digits (`12345`) or containing five digits followed by a hyphen, and then four more digits (`12345-1234`).

If all of the inputs in an application are these kinds of fixed phrases, regular expressions can do the job without requiring full-scale NLP. If the entire problem can be solved with regular expressions, then NLP isn't needed. If only part of the problem can be solved with regular expressions, but part of it needs NLP, regular expressions can be combined with natural language techniques. For example, if the text includes formatted numbers such as phone numbers, zip codes, or dates, regular expressions can be used to just analyze those numbers. Python has excellent libraries for handling regular expressions if regular expressions are needed in an application. We will discuss combining NLP and regular expressions in *Chapter 8* and *Chapter 9*.

Recognizing inputs from a known list of words

If the only available inputs are from a fixed set of possibilities, then NLP isn't needed. For example, if the input can only be a US state, then the application can just look for the names of states. Things can get a little more complicated if the inputs include words from a fixed set of possibilities, but there are surrounding words. This is called **keyword spotting**. This can happen if the desired response is from a fixed set of words, such as the name of one of 50 states, and the users sometimes add something – for example, the user says *I live in Arizona* in response to a system question like *Where do you live?*

NLP is probably not needed for this – the system just has to be able to ignore the irrelevant words (*I live in*, in this example). Regular expressions can be written to ignore irrelevant words by using **wildcard** characters. Python regular expressions use * to match any number of characters, including zero. Python uses + to match at least one character. So, a regular expression for spotting the keyword `Arizona` in Python would just be `*Arizona*`.

Using graphical interfaces

Most applications rely on a **graphical user interface**, where the user interacts with the application by selecting choices from menus and clicking buttons. These conventional interfaces are easier to build than NLU-based interfaces and are perfectly suitable for many applications. When is an NLU-based interface a better choice?

NLU is a better choice as the information that the user has to supply becomes more detailed. When this happens, a graphical interface has to rely on deeper and deeper levels of menus, requiring users to navigate through menu after menu until they find the information they need or until the application has collected enough information to answer their questions. This is especially a problem with mobile interfaces, where the amount of information that can fit on the screen is much less than the amount of information that fits on a laptop or desktop computer, which means that the menus need to have deeper levels. On the other hand, an NLU input allows the user to state their goal once, without having to navigate through multiple menus.

Another problem that graphical interfaces with deep menus have is that the terminology used in the menus does not always match the users' mental models of their goals. These mismatches can lead users down the wrong path. They might not realize their mistake until several levels farther down in the menu tree. When that happens, the user has to start all over again.

The contrast between graphical and NLP applications can easily be seen on websites and applications that include both a conventional graphical interface and an NLP chatbot. In those interfaces, the user can choose between menu-based navigation and interacting with the chatbot. A good example is the Microsoft Word 2016 interface. Word is a very complex application with a rich set of capabilities. Making an intuitive graphical interface for an application that is this complex is difficult, and it can be hard for users to find the information they're looking for.

To address this, Microsoft provides both graphical and NLP interfaces to Word functionality. At the top of the page of a Word document, there are choices including **Home**, **Insert**, **Draw**, and **Layout**. Clicking on any of these changes the ribbon to offer many more choices, often opening up more and more levels of menus. This is the graphical approach. But Word also offers a **Tell me** option as one of the top-level menu options. If the user selects **Tell me**, they can type natural language questions about how to do things in Word. For example, typing How do I add an equation will provide a list of several different ways to add an equation to a Word document. This is much quicker and more direct than looking through nested menus.

Developers should consider adding NLU functionality to graphical applications when menu levels get more than three or so levels deep, especially if each menu level has many choices.

So far, we've looked at many factors that make an application more or less suited for NLP technology. The next considerations are related to the development process – the availability of data and the development process itself, which we discuss in the next sections.

Ensuring that sufficient data is available

Having determined whether the problem is suitable for NLU, we can turn to the next question – what kinds of data are available for addressing this problem? Is there existing data? If not, what would be involved in obtaining the kind of data that's needed to solve the problem?

We will look at two kinds of data. First, we will consider *training data*, or examples of the kinds of language that will be used by users of NLU systems, and we will look at sources of training data. The second kind of data that we will discuss is *application data*. The information in this section will enable you to determine whether you have enough training data and how much work it will take to format it properly to be used in the NLU system development process.

Application data is the information that the system will use to answer users' questions. As we will see, it can come from publicly available sources or from internal databases. For application data, we will see that it is important to ensure that the data is available, reliable, and can be obtained without excessive cost.

Training data

Natural language applications are nearly all trained based on examples of the kinds of inputs they're expected to process. That means that sufficient training data needs to be available in order for any natural language application to be successful. Not having enough training data means that when the application is deployed, there will be inputs that can't be processed because the system hasn't been exposed to any similar inputs during the development phase. This doesn't mean that the system needs to see every possible input during training. This is nearly impossible, especially if the intended inputs are long or complex documents such as product reviews.

It is extremely unlikely that the same review will occur more than once. Rather, the training process is designed so that documents that are semantically similar will be analyzed in the same way, even if the exact words and phrasings are different.

Machine learning algorithms such as those we'll be learning about in *Chapter 9* and *Chapter 10*, require fairly large amounts of data. The more different categories or intents that have to be distinguished, the more data is required. Most practical applications will need thousands of training examples.

In addition to examples, normally the training data also has to include the *right answer* or how the trained system is expected to analyze the data. The technical term for the *right answer* is **annotation**. Annotations can also be referred to as the **ground truth** or **gold standard**. For example, if the application is designed to determine whether a product review is positive or negative, annotations (provided by human judges) assign a positive or negative label to a set of reviews that will be used as training and test data.

Table 2.2 shows examples of positive and negative product reviews and their annotations. An accurate system for classifying product reviews would probably need to be based on several thousand product reviews. In some cases, as in the examples in *Table 2.2*, the task of annotation doesn't require any special expertise; almost anyone with a reasonable command of English can decide whether a product review is positive or negative. This means that simple annotation tasks can be inexpensively crowdsourced.

On the other hand, some annotations have to be done by subject matter experts. For example, annotating data from an interactive troubleshooting dialog for a complex software product would probably need to be done by someone with expertise in that product. This would make the annotation process much more expensive and might not even be possible if the necessary experts aren't available:

Text	Annotation
I was very disappointed with this product. It was flimsy, overpriced, and the paint flaked off.	Negative
This product met my every expectation. It is well made, looks great, and the price is right. I have no reservations about recommending it to anyone.	Positive

Table 2.2 – Examples of positive and negative annotations of product reviews

Although data annotation can be difficult and expensive, not all NLU algorithms require annotated data. In particular, unsupervised learning, which we will cover in *Chapter 12*, is based on unannotated data. In *Chapter 12*, we will also discuss the limits of unannotated data.

The full set of training examples for an application is called a **corpus**, or **dataset**. It is essential to have sufficient training data in order for the application to be accurate. The training data does not have to be available all at once – development can begin before the data collection is complete, and additional data can be added as development progresses. This can lead to problems with consistency if annotators forget the criteria that they used to annotate earlier data.

Where does data come from? Python NLP libraries contain several toy datasets that can be used to test system setup or algorithms, or that can be used in student projects where there's no plan to put a system into production. In addition, larger datasets can also be obtained from organizations such as Hugging Face (`https://huggingface.co/`) or the Linguistic Data Consortium (`https://www.ldc.upenn.edu/`).

For enterprise applications, preexisting data from an earlier application that was performed by human agents can be very helpful. Examples of this could include transcripts of customer service calls with agents.

Another good source of data is the text fields of databases. For example, this is probably where you would expect to find product reviews for an organization's products. In many cases, text fields of databases are accompanied by another field with a manual classification that identifies, for example, whether the review is positive or negative. This manual classification is, in effect, an annotation that can be used in the training process to create a system that can automatically classify product reviews.

Finally, new data can also be collected specifically to support an application. This can be time-consuming and expensive, but sometimes it's the only way to get the appropriate data. Data collection can be a complex topic in itself, especially when the data is collected to support interactive dialogs with human users.

Data, including data collection, is discussed in more detail in *Chapter 5*.

Application data

In addition to the data required to train the natural language application, it's important to take into account any costs associated with accessing the information that the system will be providing.

Many third-party web services provide APIs that can be accessed by developers to obtain free or paid information. There are some websites that provide general information about available public APIs, such as **APIsList** (`https://apislist.com/`). This site lists APIs that can deliver data on topics ranging over hundreds of categories including weather, social networks, mapping, government, travel, and many more. Many APIs require payments, either as a subscription or per transaction, so it's important to consider these potential costs when selecting an application.

Taking development costs into account

After making sure that data is available, and that the data is (or can be) annotated with the required intents, entities, and classification categories, the next consideration for deciding whether NLP is a good fit for an application is the cost of developing the application itself. Some technically feasible applications can nevertheless be impractical because they would be too costly, risky, or time-consuming to develop.

Development costs include determining the most effective machine learning approaches to a specific problem. This can take significant time and involve some trial and error as models need to be trained and retrained in the process of exploring different algorithms. Identifying the most promising algorithms is also likely to require NLP data scientists, who may be in short supply. Developers have to ask the question of whether the cost of development is consistent with the benefits that will be realized by the final application.

For low-volume applications, it should also be kept in mind that the cost of developing and deploying an NLP solution can exceed the cost of employing humans to perform the same tasks. This is particularly true if some humans will still be needed for more complex tasks, even if an NLP solution is implemented and is doing part of the work

Taking maintenance costs into account

The final consideration for natural language applications, especially deployed applications, is the cost of maintenance. This is easy to overlook because NLU applications have several maintenance considerations that don't apply to most traditional applications. Specifically, the type of language used in some applications changes over time. This is expected since it reflects changes in the things that the users are talking about. In customer service applications, for example, product names, store locations, and services change, sometimes very quickly. The new vocabulary that customers use to ask about this information changes as well. This means that new words have to be added to the system, and machine learning models have to be retrained.

Similarly, applications that provide rapidly changing information need to be kept up to date on an ongoing basis. As an example, the word *COVID-19* was introduced in early 2020 – no one had ever heard it before, but now it is universally familiar. Since medical information about COVID-19 changes rapidly, a chatbot designed to provide COVID-19 information will have to be very carefully maintained in order to ensure that it's up to date and is not providing incorrect or even harmful information.

In order to keep applications up to date with the users' topics, three tasks that are specific to natural language applications need to be planned for:

- Developers need to be assigned to keep the application up to date as new information (such as new products or new product categories) is added to the system.

- Frequent review of platform-provided logging of user inputs should be done. User inputs that are not handled correctly must be analyzed to determine the correct way of handling them. Are the users asking about new topics (intents)? Then new intents have to be added. Are they talking about existing topics in different ways? If that's the case, new training examples need to be added to the existing intents.

- When issues are discovered and user inputs are not being handled correctly, the system needs to be modified. The simplest type of modification is adding new vocabulary, but in some cases, more structural changes are necessary. For example, it may be that an existing intent has to be split into multiple intents, which means that all the training data for the original intent has to be reviewed.

The number of developers required to keep the application updated depends on several considerations:

- **The number of user inputs**: If the system gets hundreds or thousands of failed inputs per day, developers need to be assigned to review these and add information to the system so that it can handle these inputs

- **The complexity of the application**: If the application includes hundreds of intents and entities, it will take more developers to keep it up to date and ensure that any new information stays consistent with old information

- **The volatility of the information provided by the application**: If the application is one where new words, new products, and new services are continually being added, the system will require more frequent changes to stay up to date

These costs are in addition to any costs for hardware or cloud services that are not specific to natural language applications.

A flowchart for deciding on NLU applications

This chapter has covered many considerations that should be taken into account in deciding on an NLP application.

Figure 2.2 summarizes these considerations as a flowchart of the process for evaluating a potential NLU application.

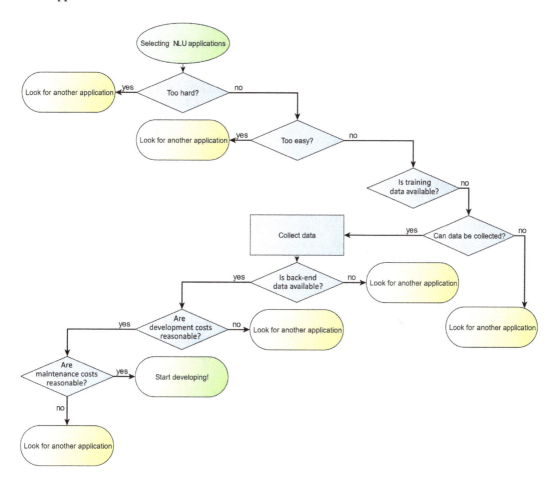

Figure 2.2 – Steps in evaluating an NLU project

Starting at the top, the process starts by asking whether the problem is too hard or too easy for the current state of the art, using the criteria discussed earlier. If it's either too hard or too easy, we should look for another application, or look at cutting back or expanding the scope of the application to make it a better fit for NLP technology. For example, the application might be redesigned to handle fewer languages.

If the problem seems to be a good fit for the state of the art, the next steps are to ensure that the appropriate data is available, and if not, whether data can be collected. Once data is available, the next thing to look at is to see whether the costs of development and maintenance are reasonable. If everything looks good, work on the application can proceed.

Summary

In this chapter, we covered the topic of selecting NLP applications that have a good chance of success with current NLP technology. Successful applications generally have input with specific, objective answers, have training data available, and handle (at most) a few languages.

Specifically, this chapter addressed a number of important questions. We learned how to identify problems that are the appropriate level of difficulty for the current state of the art of NLU technology. We also learned how to ensure that sufficient data is available for system development and how to estimate the costs of development and maintenance.

Learning how to evaluate the feasibility of different types of NLP applications as discussed in this chapter will be extremely valuable as you move forward with your NLP projects. Selecting an application that is too ambitious will result in frustration and a failed project, whereas selecting an application that is too easy for the state of the art will lead to wasted time and an unnecessarily complex system.

We have achieved our goal of learning how to evaluate the feasibility of NLP projects in terms of important criteria such as technical feasibility as well as the practical considerations of data availability and maintenance costs.

In the next chapter, we will look at the major approaches to NLP and the advantages and disadvantages of each approach. These approaches include rule-based systems, in which human experts write rules that describe how the system should analyze inputs, and machine learning, where the system is trained to analyze inputs by processing many examples of inputs and how they should be analyzed.

Part 2: Developing and Testing Natural Language Understanding Systems

After completing this section, you will be able to decide what techniques are applicable to address a problem with natural language understanding technologies and implement a system using Python and Python libraries such as NLTK, spaCy, and Keras, and evaluate it.

This part comprises the following chapters:

- *Chapter 3, Approaches to Natural Language Understanding – Rule-Based Systems, Machine Learning, and Deep Learning*
- *Chapter 4, Selecting Libraries and Tools for Natural Language Understanding*
- *Chapter 5, Natural Language Data – Finding and Preparing Data*
- *Chapter 6, Exploring and Visualizing Data*
- *Chapter 7, Selecting Approaches and Representing Data*
- *Chapter 8, Rule-Based Techniques*
- *Chapter 9, Machine Learning Part 1 – Statistical Machine Learning*
- *Chapter 10, Machine Learning Part 2 – Neural Networks and Deep Learning Techniques*
- *Chapter 11, Machine Learning Part 3 – Transformers and Large Language Models*
- *Chapter 12, Applying Unsupervised Learning Approaches*
- *Chapter 13, How Well Does It Work? – Evaluation*

3

Approaches to Natural Language Understanding – Rule-Based Systems, Machine Learning, and Deep Learning

This chapter will review the most common approaches to **natural language understanding** (NLU) and discuss both the benefits and drawbacks of each approach, including rule-based techniques, statistical techniques, and deep learning. It will also discuss popular pre-trained models such as **Bidirectional Encoder Representations from Transformers** (**BERT**) and its variants. We will learn that NLU is not a single technology; it includes a range of techniques, which are applicable to different goals.

In this chapter, we cover the following main topics:

- Rule-based approaches
- Traditional machine-learning approaches
- Deep learning approaches
- Pre-trained models
- Considerations for selecting technologies

Let's begin!

Rule-based approaches

The basic idea behind **rule-based approaches** is that language obeys rules about how words are related to their meanings. For example, when we learn foreign languages, we typically learn specific rules about what words mean, how they're ordered in sentences, and how prefixes and suffixes change the meanings of words. The rule-based approach to NLU operates on the premise that these kinds of rules can be provided to an NLU system so that the system can determine the meanings of sentences in the same way that a person does.

The rule-based approach was widely used in NLU from the mid-1950s through the mid-1990s until machine-learning-based approaches became popular. However, there are still NLU problems where rule-based approaches are useful, either on their own or when combined with other techniques.

We will begin by reviewing the rules and data that are relevant to various aspects of language.

Words and lexicons

Nearly everyone is familiar with the idea of words, which are usually defined as units of language that can be spoken individually. As we saw in *Chapter 1*, in most, but not all, languages, words are separated by white space. The set of words in a language is referred to as the language's **lexicon**. The idea of a lexicon corresponds to what we think of as a dictionary – a list of the words in a language. A computational lexicon also includes other information about each word. In particular, it includes its part or parts of speech. Depending on the language, it could also include information on whether the word has irregular forms (such as, for the irregular English verb "*eat*," the past tense "*ate*" and the past participle "*eaten*" are irregular). Some lexicons also include semantic information such as words that are related in meaning to each word.

Part-of-speech tagging

Traditional parts of speech as taught in schools include categories such as "*noun*", "*verb*", "*adjective*", *preposition*, and so on. The parts of speech used in computational lexicons are usually more detailed than these since they need to express more specific information than what could be captured by traditional categories. For example, the traditional *verb* category in English is usually broken down into several different parts of speech corresponding to the different forms of the verb, such as the past tense and past participle forms. A commonly used set of parts of speech for English is the parts of speech from the Penn Treebank (`https://catalog.ldc.upenn.edu/LDC99T42`). Different languages will have different parts of speech categories in their computational lexicons.

A very useful task in processing natural language is to assign parts of speech to the words in a text. This is called **part-of-speech tagging** (**POS tagging**). *Table 3.1* shows an example of the Penn Treebank part-of-speech tags for the sentence "*We would like to book a flight from Boston to London*:"

Word	Part of speech	Meaning of part of speech label
we	PRP	Personal pronoun
would	MD	Modal verb
like	VB	Verb, base form
to	TO	To (this word has its own part of speech)
book	VB	Verb, base form
a	DT	Determiner (article)
flight	NN	Singular noun
from	IN	Preposition
Boston	NNP	Proper noun
to	TO	To
London	NNP	Proper noun

Table 3.1 – Part-of-speech tags for "We would like to book a flight from Boston to London"

POS tagging is not just a matter of looking words up in a dictionary and labeling them with their parts of speech because many words have more than one part of speech. In our example, one of these is "*book*," which is used as a verb in the example but is also commonly used as a noun. POS tagging algorithms have to not only look at the word itself but also consider its context in order to determine the correct part of speech. In this example, "*book*" follows "*to*," which often indicates that the next word is a verb.

Grammar

Grammar rules are the rules that describe how words are ordered in sentences so that the sentences can be understood and also so that they can correctly convey the author's meaning. They can be written in the form of rules describing part-whole relationships between sentences and their components. For example, a common grammar rule for English says that a sentence consists of a noun phrase followed by a verb phrase. A full computational grammar for any natural language usually consists of hundreds of rules and is very complex. It isn't very common now to build grammar from scratch; rather, grammar is already included in commonly used Python NLP libraries such as **natural language toolkit (NLTK)** and **spaCy**.

Parsing

Finding the relationships between parts of a sentence is known as **parsing**. This involves applying the grammar rules to a specific sentence to show how the parts of the sentence are related to each other. *Figure 3.1* shows the parsing of the sentence "*We would like to book a flight.*" In this style of parsing, known as **dependency parsing**, the relationships between the words are shown as arcs between the words. For example, the fact that "*we*" is the subject of the verb "*like*" is shown by an arc labeled **nsubj** connecting "*like*" to "*we*".

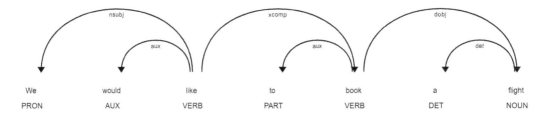

Figure 3.1 – Parsing for "We would like to book a flight"

At this point, it isn't necessary to worry about the details of parsing – we will discuss it in more detail in *Chapter 8*.

Semantic analysis

Parsing involves determining how words are structurally related to each other in sentences, but it doesn't say anything about their meanings or how their meanings are related. This kind of processing is done through **semantic analysis**. There are many approaches to semantic analysis—this is an active research field—but one way to think of semantic analysis is that it starts with the main verb of the sentence and looks at the relationships between the verb and other parts of the sentence, such as the subject, direct object, and related prepositional phrases. For example, the subject of "*like*" in *Figure 3.1* is "*We.*" "*We*" could be described as the "*experiencer*" of "*like*" since it is described as experiencing "*liking.*" Similarly, the thing that is liked, "*to book a flight*," could be described as the "*patient*" of "*like.*" Semantic analysis is most frequently done through the application of rules, but it can also be done with machine learning techniques, as described in the next sections.

Finding semantic relationships between the concepts denoted by words, independently of their roles in sentences, can also be useful. For example, we can think of a "*dog*" as a kind of "*animal*," or we can think of "*eating*" as a kind of action. One helpful resource for finding these kinds of relationships is Wordnet (https://wordnet.princeton.edu/), which is a large, manually prepared database describing the relationships between thousands of English words. *Figure 3.2* shows part of the Wordnet information for the word "*airplane*," indicating that an airplane is a kind of "*heavier-than-aircraft*," which is a kind of "*aircraft*," and so on, going all the way up to the very general category "*entity*."

- <u>S:</u> (n) **airplane**, *aeroplane*, *plane*
  - *direct hyponym* / *full hyponym*
  - *part meronym*
  - *domain term category*
  - *direct hyponym* / *inherited hyponym* / *sister term*
    - <u>S:</u> (n) heavier-than-air craft
      - <u>S:</u> (n) aircraft
        - <u>S:</u> (n) craft
          - <u>S:</u> (n) vehicle
            - <u>S:</u> (n) conveyance, transport
              - <u>S:</u> (n) instrumentality, instrumentation
                - <u>S:</u> (n) artifact, artefact
                  - <u>S:</u> (n) whole, unit
                    - <u>S:</u> (n) object, physical object
                      - <u>S:</u> (n) physical entity
                        - <u>S:</u> (n) entity

Figure 3.2 – Wordnet semantic hierarchy for the word "airplane"

Pragmatic analysis

Pragmatic analysis determines the meanings of words and phrases in context. For example, in long texts, different words can be used to refer to the same thing, or different things can be referred to with the same word. This is called **coreference**. For example, the sentence "*We want to book a flight from Boston to London*" could be followed by "*the flight needs to leave before 10 a.m.*" Pragmatic analysis determines that the flight that needs to leave before 10 a.m. is the same flight that we want to book.

A very important type of pragmatic analysis is **named entity recognition** (**NER**), which links up references that occur in texts to corresponding entities in the real world. *Figure 3.3* shows NER for the sentence "*Book a flight to London on United for less than 1,000 dollars.*" "*London*" is a named entity, which is labeled as a geographical location, "*United*" is labeled as an organization, and "*less than 1,000 dollars*" is labeled as a monetary amount:

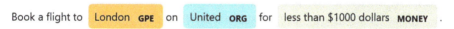

Figure 3.3 – NER for "Book a flight to London on United for less than 1,000 dollars"

Pipelines

In NLP applications, the steps we have just described are most often implemented as a **pipeline**; that is, a sequence of steps where the results of one step are the input to the next step. For example, a typical NLP pipeline might be as follows:

- **Lexical lookup**: Look up the words in the application's dictionary

- **POS tagging**: Assign parts of speech to each word in context

- **Parsing**: Determine how the words are related to each other

- **Semantic analysis**: Determine the meanings of the words and the overall meaning of the sentence

- **Pragmatic analysis**: Determine aspects of the meaning that depend on a broader context, such as the interpretations of pronouns

One advantage of using pipelines is that each step can be implemented with different technologies, as long as the output of that step is in the format expected by the next step. So, pipelines are not only useful in rule-based approaches but also in the other techniques, which we will describe in the next sections.

More details on rule-based techniques will be provided in *Chapter 8*.

We will now turn to techniques that rely less on the rules of the language and more on machine learning with existing data.

Traditional machine learning approaches

While rule-based approaches provide very fine-grained and specific information about language, there are some drawbacks to these approaches, which has motivated the development of alternatives. There are two major drawbacks:

- Developing the rules used in rule-based approaches can be a laborious process. Rule development can either be done by experts directly writing rules based on their knowledge of the language or, more commonly, the rules can be derived from examples of text that have been annotated with a correct analysis. Both of these approaches can be expensive and time-consuming.

- Rules are not likely to be universally applicable to every text that the system encounters. The experts who developed the rules might have overlooked some cases, the annotated data might not have examples of every case, and speakers can make errors such as false starts, which need to be analyzed although they aren't covered by any rule. Written language can include spelling errors, which results in words that aren't in the lexicon. Finally, the language itself can change, resulting in new words and new phrases that aren't covered by existing rules.

For these reasons, rule-based approaches are primarily used as part of NLU pipelines, supplementing other techniques.

Traditional machine learning approaches were motivated by problems in classification, where documents that are similar in meaning can be grouped. Two problems have to be solved in classification:

- Representing the documents in a training set in such a way that documents in the same categories have similar representations

- Deciding how new, previously unseen documents should be classified based on their similarity to the documents in the training set

Representing documents

Representations of documents are based on words. A very simple approach is to assume that the document should be represented simply as the set of words that it contains. This is called the **bag of words (BoW)** approach. The simplest way of using a BoW for document representation is to make a list of all the words in the corpus, and for each document and each word, state whether that word occurs in that document.

For example, suppose we have a corpus consisting of the three documents in *Figure 3.4*:

1. I'm looking for a nearby Chinese restaurant that's highly rated.

2. An Italian restaurant within five miles of here.

3. Are there any inexpensive Middle Eastern places that aren't too far away?

Figure 3.4 – A small corpus of restaurant queries

The entire vocabulary for this toy corpus is 29 words, so each document is associated with a list 29 items long that states, for each word, whether or not it appears in the document. The list represents *occurs* as 1 and *does not occur* as 0, as shown in *Table 3.2*:

	a	an	any	are	aren't	away	Chinese	Eastern	...
1	1	0	0	0	0	0	1	0	...
2	0	1	0	0	0	0	0	0	...
3	0	0	1	1	1	1	0	1	...

Table 3.2 – BoW for a small corpus

Table 3.2 shows the BoW lists for the first eight words in the vocabulary for these three documents. Each row in *Table 3.2* represents one document. For example, the word "*a*" occurs once in the first document, but the word "*an*" does not occur, so its entry is *0*. This representation is mathematically a *vector*. Vectors are a powerful tool in NLU, and we will discuss them later in much more detail.

The BoW representation might seem very simplistic (for example, it doesn't take into account any information about the order of words). However, other variations on this concept are more powerful, which will be discussed later on in *Chapters 9* to *12*.

Classification

The assumption behind the BoW approach is that the more words that two documents have in common, the more similar they are in meaning. This is not a hard-and-fast rule, but it turns out to be useful in practice.

For many applications, we want to group documents that are similar in meaning into different categories. This is the process of **classification**. If we want to classify a new document into one of these categories, we need to find out how similar its vector is to the vectors of the other documents in each category. For example, the sentiment analysis task discussed in *Chapter 1* is the task of classifying documents into one of two categories – positive or negative sentiment regarding the topic of the text.

Many algorithms have been used for the classification of text documents. Naïve Bayes and **support vector machines** (**SVMs**), to be discussed in *Chapter 9*, are two of the most popular. Neural networks, especially **recurrent neural networks** (**RNNs**), are also popular. Neural networks are briefly discussed in the next section and will be discussed in detail in *Chapter 10*.

In this section, we've summarized some approaches that are used in traditional machine learning. Now, we will turn our attention to the new approaches based on deep learning.

Deep learning approaches

Neural networks, and especially the large neural networks generally referred to as **deep learning**, have become very popular for NLU in the past few years because they significantly improve the accuracy of earlier methods.

The basic concept behind neural networks is that they consist of layers of connected units, called **neurons** in analogy to the neurons in animal nervous systems. Each neuron in a neural net is connected to other neurons in the neural net. If a neuron receives the appropriate inputs from other neurons, it will fire, or send input to another neuron, which will in turn fire or not fire depending on other inputs that it receives. During the training process, weights on the neurons are adjusted to maximize classification accuracy.

Figure 3.5 shows an example of a four-layer neural net performing a sentiment analysis task. The neurons are circles connected by lines. The first layer, on the left, receives a text input. Two hidden layers of neurons then process the input, and the result (*positive*) is produced by the single neuron in the final output layer:

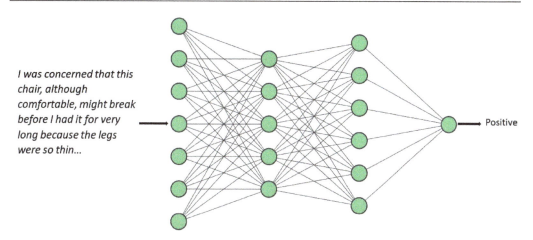

Figure 3.5 – A four-layer neural network trained to perform sentiment analysis on product reviews

Although the concepts behind neural networks have existed for many years, the implementation of neural networks large enough to perform significant tasks has only been possible within the last few years because of the limitations of earlier computing resources. Their current popularity is largely due to the fact that they are often more accurate than earlier approaches, especially given sufficient training data. However, the process of training a neural net for large-scale tasks can be complex and time-consuming and can require the services of expert data scientists. In some cases, the additional accuracy that a neural net provides is not enough to justify the additional expense of developing the system.

Deep learning and neural networks will be discussed in detail in *Chapter 10*.

Pre-trained models

The most recent approach to NLU is based on the idea that much of the information required to understand natural language can be made available to many different applications by processing generic text (such as internet text) to create a baseline model for the language. Some of these models are very large and are based on tremendous amounts of data. To apply these models to a specific application, the generic model is adapted to the application through the use of application-specific training data, through a process called **fine-tuning**. Because the baseline model already contains a vast amount of general information about the language, the amount of training data can be considerably less than the training data required for some of the traditional approaches. These popular technologies include BERT and its many variations and **Generative Pre-trained Transformers** (**GPTs**) and their variations.

Pre-trained models will be discussed in detail in *Chapter 11*.

Considerations for selecting technologies

This chapter has introduced four classes of NLU technologies:

- Rule-based
- Statistical machine learning
- Deep learning and neural networks
- Pre-trained models

How should we decide which technology or technologies should be employed to solve a specific problem? The considerations are largely practical and have to do with the costs and effort required to create a working solution. Let's look at the characteristics of each approach.

Table 3.3 lists the four approaches to NLU that we've reviewed in this chapter and how they compare with respect to developer expertise, the amount of data required, the training time, accuracy, and cost. As *Table 3.3* shows, every approach has advantages and disadvantages. For small or simple problems that don't require large amounts of data, the rule-based, deep learning, or pre-trained approaches should be strongly considered, at least for part of the pipeline. While pre-trained models are accurate and have relatively low development costs, developers may prefer to avoid the costs of cloud services or the costs of managing large models on their local computer resources:

	Developer expertise	Amount of data required	Training time	Accuracy	Cost
Rule-based	High (linguists or domain experts)	Small amount of domain-specific data	Large amount of time for experts to write rules	High if rules are accurate	Rule development may be costly; computer time costs are low
Statistical	Medium – use standard tools; some NLP/data science expertise required	Medium amount of domain-specific data	Large amount of time for annotation	Medium	Data annotation may be costly; computer time costs are low

	Developer expertise	Amount of data required	Training time	Accuracy	Cost
Deep learning	High (data scientists)	Large amount of domain-specific data	Large amount of time for annotation; additional computer time for training models	Medium-high	Charges for some cloud services or local computer resources
Pre-trained	Medium – use standard tools with some data science expertise	Small amount of domain-specific data	Medium amount of time to label data for fine-tuning models	High	Charges for some cloud services or local computer resources

Table 3.3 – Comparison between general approaches to NLU

The most important considerations are the problem that's being addressed and what the acceptable costs are. It should also be kept in mind that choosing one or another technology isn't a permanent commitment, especially for approaches that rely on annotated data, which can be used for more than one approach.

Summary

In this chapter, we surveyed the various techniques that can be used in NLU applications and learned several important skills.

We learned about what rule-based approaches are and the major rule-based techniques, including topics such as POS tagging and parsing. We then learned about the important traditional machine learning techniques, especially the ways that text documents can be represented numerically. Next, we focused on the benefits and drawbacks of the more modern deep learning techniques and the advantages of pre-trained models.

In the next chapter, we will review the basics of getting started with NLU – installing Python, using Jupyter Labs and GitHub, using NLU libraries such as NLTK and spaCy, and how to choose between libraries.

4
Selecting Libraries and Tools for Natural Language Understanding

This chapter will get you set up to process natural language. We will begin by discussing how to install Python, and then we will discuss general software development tools such as JupyterLab and GitHub. We will also review major Python **natural language processing** (**NLP**) libraries, including the **Natural Language Toolkit** (**NLTK**), **spaCy**, and **TensorFlow/Keras**.

Natural language understanding (**NLU**) technology has benefited from a wide assortment of very capable, freely available tools. While these tools are very powerful, there is no one library that can do all of the NLP tasks needed for all applications, so it is important to understand what the strengths of the different libraries are and how to combine them.

Making the best use of these tools will greatly accelerate any NLU development project. These tools include the Python language itself, development tools such as JupyterLab, and a number of specific natural language libraries that can perform many NLU tasks. It is equally important to know that because these tools are widely used by many developers, active online communities such as Stack Overflow (https://stackoverflow.com/) have developed. These are great resources for getting answers to specific technical questions.

This chapter will cover the following topics:

- Installing Python
- Developing software—JupyterLab and GitHub
- Exploring the libraries
- Looking at an example

Since there are many online resources for using tools such as Python, JupyterLab, and GitHub, we will only briefly outline their usage here in order to be able to spend more time on NLP.

> **Note**
>
> For simplicity, we will illustrate the installation of the libraries in the base system. However, you may wish to install the libraries in a virtual environment, especially if you are working on several different Python projects. The following link may be helpful for installing a virtual environment: `https://packaging.python.org/en/latest/guides/installing-using-pip-and-virtual-environments/`.

Technical requirements

To run the examples in this chapter, you will need the following software:

- Python 3
- `pip` or `conda` (preferably `pip`)
- JupyterLab
- NLTK
- spaCy
- Keras

The next sections will go over the process of installing these packages, which should be installed in the order in which they are listed here.

Installing Python

The first step in setting up your development environment is to install Python. If you have already installed Python on your system, you can skip to the next section, but do make sure that your Python installation includes Python 3, which is required by most NLP libraries. You can check your Python version by entering the following command in a command-line window, and the version will be displayed:

```
$ python --version
```

Note that if you have both Python 2 and Python 3 installed, you may have to run the `python3 --version` command to check the Python 3 version. If you don't have Python 3, you'll need to install it. Some NLP libraries require not just Python 3 but Python 3.7 or greater, so if your version of Python is older than 3.7, you'll need to update it.

Python runs on almost any operating system that you choose to use, including Windows, macOS, and Linux. Python can be downloaded for your operating system from `http://www.python.org`.

Download the executable installer for your operating system and run the installer. When Python is installed, you can check the installation by running the preceding command on your command line or terminal. You will see the version you've just installed, as shown in the following command-line output:

```
$ python --version
Python 3.8.5
```

This installs Python, but you will also need to install add-on libraries for NLP. Installing libraries is done with the auxiliary programs `pip` and `conda`.

`pip` and `conda` are two cross-platform tools that can be used for installing Python libraries. We will be using them to install several important natural language and **machine learning** (**ML**) libraries. We will primarily use `pip` in this book, but you can also use `conda` if it's your preferred Python management tool. `pip` is included by default with Python versions 3.4 and newer, and since you'll need 3.7 for the NLP libraries, `pip` should be available in your Python environment. You can check the version with the following command:

```
$ pip --version
```

You should see the following output:

```
$ pip 21.3.1 from c:\<installation dir>\pip (python 3.9)
```

In the next section, we will discuss the development environment we will be using: JupyterLab.

Developing software – JupyterLab and GitHub

The development environment can make all the difference in the efficiency of the development process. In this section, we will discuss two popular development resources: JupyterLab and GitHub. If you are familiar with other Python **interactive development environments** (**IDEs**), then you can go ahead and use the tools that you're familiar with. However, the examples discussed in this book will be shown in a JupyterLab environment.

JupyterLab

JupyterLab is a cross-platform coding environment that makes it easy to experiment with different tools and techniques without requiring a lot of setup time. It operates in a browser environment but doesn't require a cloud server—a local server is sufficient.

Installing JupyterLab is done with the following `pip` command:

```
$ pip install jupyterlab
```

Once JupyterLab is installed, you can run it using the following command:

```
$ jupyter lab
```

This command should be run in a command line in the directory where you would like to keep your code. The command will launch a local server, and the Jupyter environment will appear in a browser window, as shown in *Figure 4.1*:

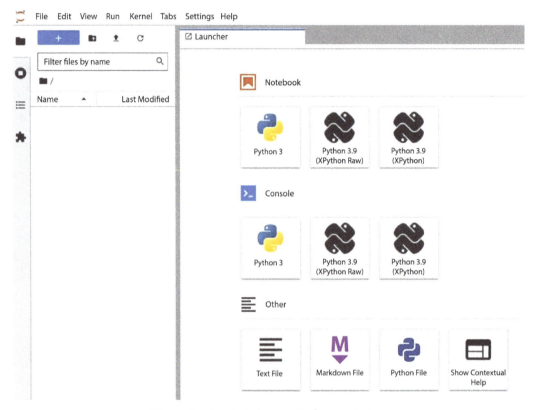

Figure 4.1 – JupyterLab user interface on startup

The environment shown in *Figure 4.1* includes three types of content, as follows:

- **Notebook**—Contains your coding projects
- **Console**—Gives you access to command-line or terminal functions from directly within the Jupyter notebook
- **Other**—Refers to other types of files that may be included in the local directory where the start command was run

When you click on the **Python 3** icon under **Notebook**, you'll get a new notebook showing a coding cell, and you'll be ready to start coding in Python. We'll return to the JupyterLab environment and start coding in Python in the *Looking at an example* section later in this chapter and again in *Chapter 5*.

GitHub

Many of you are probably already familiar with GitHub, a popular open source code repository system (`https://github.com`). GitHub provides very extensive capabilities for storing and sharing code, developing code branches, and documenting code. The core features of GitHub are currently free.

The code examples used in this book can be found at `https://github.com/PacktPublishing/Natural-Language-Understanding-with-Python`.

The next step is to learn about several important libraries, including NLTK, spaCy, and Keras, which we will be using extensively in the following chapters.

Exploring the libraries

In this section, we will review several of the major Python libraries that are used in NLP; specifically, NLTK, spaCy, and Keras. These are very useful libraries, and they can perform most basic NLP tasks. However, as you gain experience with NLP, you will also find additional NLP libraries that may be appropriate for specific tasks as well, and you are encouraged to explore those.

Using NLTK

NLTK (`https://www.nltk.org/`) is a very popular open source Python library that greatly reduces the effort involved in developing natural language applications by providing support for many frequently performed tasks. NLTK also includes many corpora (sets of ready-to-use natural language texts) that can be used for exploring NLP problems and testing algorithms.

In this section, we will go over what NLTK can do, and then discuss the NLTK installation process.

As we discussed in *Chapter 3*, many distinct tasks can be performed in an NLU pipeline as the processing moves from raw words to a final determination of the meaning of a document. NLTK can perform many of these tasks. Most of these functions don't provide results that are directly useful in themselves, but they can be very helpful as part of a pipeline.

Some of the basic tasks that are needed in nearly all natural language projects can easily be done with NLTK. For example, texts to be processed need to be broken down into words before processing. We can do this with NLTK's `word_tokenize` function, as shown in the following code snippet:

```
import nltk
import string
from nltk import word_tokenize
text = "we'd like to book a flight from boston to London"
tokenized_text = word_tokenize(text)
print(tokenized_text)
```

The result will be an array of words:

```
['we',
 "'d",
 'like',
 'to',
 'book',
 'a',
 'flight',
 'from',
 'boston',
 'to',
'London']
```

Note that the word we'd is separated into two components, we and 'd, because it is a contraction that actually represents two words: *we* and *would*.

NLTK also provides some functions for basic statistics such as counting word frequencies in a text. For example, continuing from the text we just looked at, *we'd like to book a flight from Boston to London*, we can use the NLTK FreqDist() function to count how often each word occurs:

```
from nltk.probability import FreqDist
FreqDist(tokenized_text)
FreqDist({'to': 2, 'we': 1, "'d": 1, 'like': 1, 'book': 1, 'a': 1,
'flight': 1, 'from': 1, 'boston': 1, 'london': 1})
```

In this example, we imported the FreqDist() function from NLTK's probability package and used it to count the frequencies of each word in the text. The result is a Python dict where the keys are the words and the values are how often the words occur. The word to occurs twice, and each of the other words occurs once. For such a short text, the frequency distribution is not particularly insightful, but it can be very helpful when you're looking at larger amounts of data. We will see the frequency distribution for a large corpus in the *Looking at an example* section later in this chapter.

NLTK can also do **part-of-speech** (**POS**) tagging, which was discussed in *Chapter 3*. Continuing with our example, the nltk.pos_tag(tokenized_text) function is used for POS tagging:

```
nltk.pos_tag(tokenized_text)
[('we', 'PRP'),
 ("'d", 'MD'),
 ('like', 'VB'),
 ('to', 'TO'),
 ('book', 'NN'),
 ('a', 'DT'),
 ('flight', 'NN'),
 ('from', 'IN'),
 ('boston', 'NN'),
```

```
('to', 'TO'),
('london', 'VB')]
```

Similarly, NLTK provides functions for parsing texts. Recall that parsing was discussed in *Chapter 3*. As discussed in *Chapter 2*, NLTK also provides functions for creating and applying **regular expressions (regexes)**.

These are some of the most useful capabilities of NLTK. The full set of NLTK capabilities is too large to list here, but we will be reviewing some of these other capabilities in *Chapter 6* and *Chapter 8*.

Installing NLTK

NLTK requires Python 3.7 or greater. The installation process for Windows is to run the following command in a command window:

```
$ pip install nltk
```

For a Mac or Unix environment, run the following command in a terminal window:

```
$ pip install --user -U nltk
```

In the next section, we will go over another popular NLU library, spaCy, and explain what it can do. As with NLTK, we will be using spaCy extensively in later chapters.

Using spaCy

spaCy is another very popular package that can do many of the same NLP tasks as NLTK. Both toolkits are very capable. spaCy is generally faster, and so is more suitable for deployed applications. Both toolkits support many languages, but not all NLU tasks are supported for all languages, so in making a choice between NLTK and spaCy, it is important to consider the specific language requirements for that application.

As with NLTK, spaCy can perform many basic text-processing functions. Let's check it out!

The code to set up tokenization in spaCy is very similar to the code for NLTK, with a slightly different function name. The result is an array of words, where each element is one token. Note that the `nlp` object is initialized with an `en_core_web_sm` model that tells it to use the statistics from a particular set of web-based data, `en_core_web_sm`:

```
import spacy
from spacy.lang.en import English
nlp = spacy.load('en_core_web_sm')
text = "we'd like to book a flight from boston to london"
doc = nlp(text)
print ([token.text for token in doc])
['we', "'d", 'like', 'to', 'book', 'a', 'flight', 'from', 'boston',
'to', 'london']
```

We can also calculate statistics such as the frequency of the words that occur in the text:

```
from collections import Counter
word_freq = Counter(words)
print(word_freq)
Counter({'to': 2, 'we': 1, "'d": 1, 'like': 1, 'book': 1, 'a': 1,
'flight': 1, 'from': 1, 'boston': 1, 'london': 1})
```

The only difference between spaCy and NLTK is that NLTK uses the `FreqDist` function and spaCy uses the `Counter` function. The result, a Python dict with the words as keys and the frequencies as values, is the same for both libraries.

Just as with NLTK, we can perform POS tagging with spaCy:

```
for token in doc:
    print(token.text, token.pos_)
```

This results in the following POS assignments:

```
we PRON
'd AUX
like VERB
to PART
book VERB
a DET
flight NOUN
from ADP
boston PROPN
to ADP
london PROPN
```

Unfortunately, NLTK and spaCy use different labels for the different parts of speech. This is not necessarily a problem, because there is no *correct* or *standard* set of parts of speech, even for one language. However, it's important for the parts of speech to be consistent within an application, so developers should be aware of this difference and be sure not to confuse the NLTK and spaCy parts of speech.

Another very useful capability that spaCy has is **named entity recognition** (**NER**). NER is the task of identifying references to specific persons, organizations, locations, or other entities that occur in a text. NER can be either an end in itself or it can be part of another task. For example, a company might be interested in finding when their products are mentioned on Facebook, so NER for their products would be all that they need. On the other hand, a company might be interested in finding out if their products are mentioned in a positive or negative way, so in that case, they would want to perform both NER and **sentiment analysis** (**SA**).

NER can be performed in most NLP libraries; however, it is particularly easy to do in spaCy. Given a document, we just have to request rendering of the document using the ent style, as follows:

```
import spacy
nlp = spacy.load("en_core_web_sm")
text = "we'd like to book a flight from boston to new york"
doc = nlp(text)
displacy.render(doc,style='ent',jupyter=True,options={'distance':200})
```

The rendered result shows that the boston and new york named entities are assigned a **geopolitical entity (GPE)** label, as shown in *Figure 4.2*:

we'd like to book a flight from   boston **GPE** to new york **GPE**

Figure 4.2 – NER for "we'd like to book a flight from Boston to New York"

Parsing or the analysis of syntactic relationships among the words in a sentence can be done very easily with almost the same code, just by changing the value of the style parameter to dep from ent. We'll see an example of a syntactic parse later on in *Figure 4.6*:

```
nlp = spacy.load('en_core_web_sm')
doc = nlp('they get in an accident')
displacy.render(doc,style='dep',jupyter=True,options={'distance':200})
```

Installing spaCy is done with the following pip command:

```
$ pip install -U spacy
```

The next library we will look at is the Keras ML library.

Using Keras

Keras (https://keras.io/) is another popular Python NLP library. Keras is much more focused on ML than NLTK or spaCy and will be the go-to library for NLP **deep learning (DL)** applications in this book. It's built on top of another package called TensorFlow (https://www.tensorflow.org/), which was developed by Google. Because Keras is built on TensorFlow, TensorFlow functions can be used in Keras.

Since Keras focuses on ML, it has limited capabilities for preprocessing text. For example, unlike NLTK or spaCy, it does not support POS tagging or parsing directly. If these capabilities are needed, then it's best to preprocess the text with NLTK or spaCy. Keras does support tokenization and removal of extraneous tokens such as punctuation and HTML markup.

Keras is especially strong for text-processing applications using **neural networks** (**NN**). This will be discussed in much more detail in *Chapter 10*. Although Keras includes few high-level functions for performing NLP functions such as POS tagging or parsing in one step, it does include capabilities for training POS taggers from a dataset and then deploying the tagger in an application.

Since Keras is included in TensorFlow, Keras is automatically installed when TensorFlow is installed. It is not necessary to install Keras as an additional step. Thus, the following command is sufficient:

```
$ pip install tensorflow
```

Learning about other NLP libraries

There are quite a few other Python libraries that include NLP capabilities and that can be useful in some cases. These include PyTorch (`https://pytorch.org/`) for processing based on **deep neural networks** (**DNN**), scikit-learn (`https://scikit-learn.org/stable/`), which includes general ML functions, and Gensim (`https://radimrehurek.com/gensim/`), for topic modeling, among others. However, I would recommend working with the basic packages that we've covered here for a few projects at first until you get more familiar with NLP. If you later have a requirement for additional functionality, a different language, or faster processing speed than what the basic packages provide, you can explore some of these other packages at that time.

In the next topic, we will discuss how to choose among NLP libraries. It's good to keep in mind that choosing libraries isn't an all-or-none process—libraries can easily be mixed and matched if one library has strengths that another one doesn't.

Choosing among NLP libraries

The libraries discussed in the preceding sections are all very useful and powerful. In some cases, they have overlapping capabilities. This raises the question of selecting which libraries to use in a particular application. Although all of the libraries can be combined in the same application, it reduces the complexity of applications if fewer libraries are used.

NLTK is very strong in corpus statistics and rule-based linguistic preprocessing. For example, some useful corpus statistics include counting words, counting parts of speech, counting pairs of words (bigrams), and tabulating words in context (concordances). spaCy is fast, and its displaCy visualization library is very helpful in gaining insight into processing results. Keras is very strong in DL.

During the lifetime of a project, it is often useful to start with tools that help you quickly get a good overall picture of the data, such as NLTK and spaCy. This initial analysis will be very helpful for selecting the tools that are needed for full-scale processing and deployment. Since training DL models using tools such as Keras can be very time-consuming, doing some preliminary investigation with more traditional approaches will help narrow down the possibilities that need to be investigated in order to select a DL approach.

Learning about other packages useful for NLP

In addition to packages that directly support NLP, there are also a number of other useful general-purpose open source Python packages that provide tools for generally managing data, including natural language data. These include the following:

- **NumPy**: NumPy (`https://numpy.org/`) is a powerful package that includes many functions for the numerical calculations that we'll be working with in *Chapter 9*, *Chapter 10*, *Chapter 11*, and *Chapter 12*

- **pandas**: pandas (`https://pandas.pydata.org/`) provides general tools for data analysis and manipulation, including natural language data, especially data in the form of tables

- **scikit-learn**: scikit-learn is a powerful package for ML, including text processing (`https://scikit-learn.org/stable/`)

There are also several visualization packages that will be very helpful for graphical representations of data and for processing results. Visualization is important in NLP development because it can often give you a much more comprehensible representation of results than a numerical table. For example, visualization can help you see trends, pinpoint errors, and compare experimental conditions. We'll be using visualization tools throughout the book, but especially in *Chapter 6*. Visualization tools include generic tools for representing different kinds of numerical results, whether they have to do with NLP or not, as well as tools specifically designed to represent natural language information such as parses and NER results. Visualization tools include the following:

- **Matplotlib**: Matplotlib (`https://matplotlib.org/`) is a popular Python visualization library that's especially good at creating plots of data, including NLP data. If you're trying to compare the results of processing with several different techniques, plotting the results can often provide insights very quickly about how well the different techniques are working, which can be helpful for evaluation. We will be returning to the topic of evaluation in *Chapter 13*.

- **Seaborn**: Seaborn (`https://seaborn.pydata.org/`) is based on Matplotlib. It enables developers to produce attractive graphs representing statistical information.

- **displaCy**: displaCy is part of the spaCy tools, and is especially good at representing natural language results such as POS tags, parses, and named entities, which we discussed in *Chapter 3*.

- **WordCloud**: WordCloud (`https://amueller.github.io/word_cloud/`) is a specialized library for visualizing word frequencies in a corpus, which can be useful when word frequencies are of interest. We'll see an example of a word cloud in the next section.

Up to this point, we've reviewed the technical requirements for our software development environment as well as the NLP libraries that we'll be working with. In the next section, we'll put everything together with an example.

Looking at an example

To illustrate some of these concepts, we'll work through an example using JupyterLab where we explore an SA task for movie reviews. We'll look at how we can apply the NLTK and spaCy packages to get some ideas about what the data is like, which will help us plan further processing.

The corpus (or dataset) that we'll be looking at is a popular set of 2,000 movie reviews, classified as to whether the writer expressed a positive or negative sentiment about the movie (http://www.cs.cornell.edu/people/pabo/movie-review-data/).

> **Dataset citation**
>
> *Bo Pang* and *Lillian Lee, Seeing stars: Exploiting class relationships for sentiment categorization with respect to rating scales, Proceedings of the ACL, 2005.*

This is a good example of the task of SA, which was introduced in *Chapter 1*.

Setting up JupyterLab

We'll be working with JupyterLab, so let's start it up. As we saw earlier, you can start JupyterLab by simply typing the following command into a command (Windows) or terminal (Mac) window:

```
$ jupyter lab
```

This will start a local web server and open a JupyterLab window in a web browser. In the JupyterLab window, open a new notebook by selecting **File | New | Notebook**, and an untitled notebook will appear (you can rename it at any time by selecting **File | Rename Notebook**).

We'll start by importing the libraries that we'll be using in this example, as shown next. We'll be using the NLTK and spaCy NLP libraries, as well as some general-purpose libraries for numerical operations and visualization. We'll see how these are used as we go through the example:

```
# NLP imports
import nltk
import spacy
from spacy import displacy

# general numerical and visualization imports
import pandas as pd
import seaborn as sns
```

```
import matplotlib.pyplot as plt
from collections import Counter
import numpy as np
```

Enter the preceding code into a JupyterLab cell and run the cell. Running this cell (**Run | Run Selected Cells**) will import the libraries and give you a new code cell.

Download the movie review data by typing `nltk.download()` into the new code cell. This will open a new **NLTK Downloader** window, as shown in *Figure 4.3*:

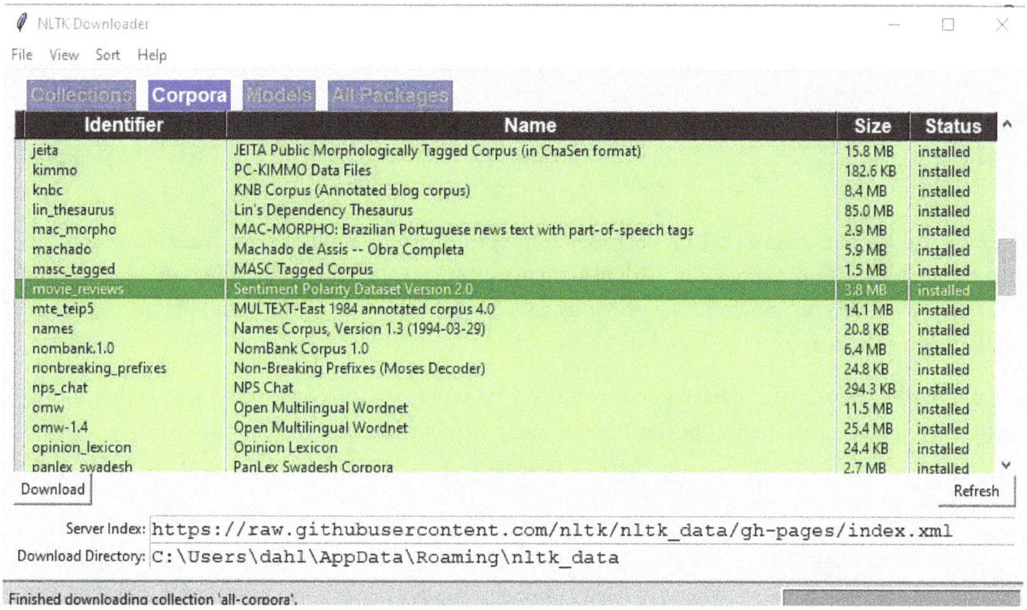

Figure 4.3 – NLTK Downloader window

On the **Corpora** tab of the **Download** window, you can select `movie_reviews`. Click the **Download** button to download the corpus. You can select **File | Change Download Directory** if you want to change where the data is downloaded. Click **File | Exit** from the downloader window to exit from the **Download** window and return to the JupyterLab interface.

If you take a look at the directory where you downloaded the data, you will see two directories: `neg` and `pos`. The directories contain negative and positive reviews, respectively. This represents the annotation of the reviews; that is, a human annotator's opinion of whether the review was positive or negative. This directory structure is a common approach for representing text classification annotations, and you'll see it in many datasets. The README file in the `movie_reviews` folder explains some details on how the annotation was done.

If you look at some of the movie reviews in the corpus, you'll see that the correct annotation for a text is not always obvious.

The next code block shows importing the reviews and printing one sentence from the corpus:

```
#import the training data
from nltk.corpus import movie_reviews
sents = movie_reviews.sents()
print(sents)
[['plot', ':', 'two', 'teen', 'couples', 'go', 'to', 'a', 'church',
'party', ',', 'drink', 'and', 'then', 'drive', '.'], ['they', 'get',
'into', 'an', 'accident', '.'], ...]
In [5]:
sample = sents[9]
print(sample)
['they', 'seem', 'to', 'have', 'taken', 'this', 'pretty', 'neat',
'concept', ',', 'but', 'executed', 'it', 'terribly', '.']
```

Since movie_reviews is an NLTK corpus, a number of corpora methods are available, including listing the sentences as an array of individual sentences. We can also select individual sentences from the corpus by number, as shown in the previous code block, where we selected and printed sentence number nine in the corpus.

You can see that the sentences have been tokenized, or separated into individual words (including punctuation marks). This is an important preparatory step in nearly all NLP applications.

Processing one sentence

Now, let's do some actual NLP processing for this sample sentence. We'll use the spaCy library to perform POS tagging and rule-based parsing and then visualize the results with the displaCy library.

We first need to create an nlp object based on web data, en_core_web_sm, which is a basic small English model. There are larger models available, but they take longer to load, so we will stick with the small model here for brevity. Then, we use the nlp object to identify the parts of speech and parse this sentence, as shown in the following code block:

```
nlp = spacy.load('en_core_web_sm')
doc = nlp('they get in an accident')
displacy.render(doc,style='dep',jupyter=True,options={'distance':200})
```

In the displacy.render command, we have requested a dependency parse (styles='dep'). This is a type of analysis we'll go into in more detail in *Chapter 8*. For now, it's enough to say that it's one common approach to showing how the words in a sentence are related to each other. The resulting dependency parse is shown in *Figure 4.4*:

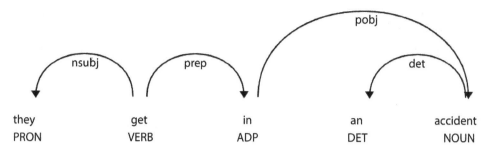

Figure 4.4 – Dependency parse for "they get in an accident"

Now that we've loaded the corpus and looked at a few examples of the kinds of sentences it contains, we will look at some of the overall properties of the corpus.

Looking at corpus properties

While corpora have many properties, some of the most interesting and insightful properties are word frequencies and POS frequencies, which we will be reviewing in the next two sections. Unfortunately, there isn't space to explore additional corpus properties in detail, but looking at word and POS frequencies should get you started.

Word frequencies

In this section, we will look at some properties of the full corpus. For example, we can look at the most frequent words using the following code:

```
words = movie_reviews.words()
word_counts = nltk.FreqDist(word.lower() for word in words if word.
isalpha())
top_words = word_counts.most_common(25)
all_fdist = pd.Series(dict(top_words))

# Setting fig and ax into variables
fig, ax = plt.subplots(figsize=(10,10))

# Plot with Seaborn plotting tools
plt.xticks(rotation = 70)
plt.title("Frequency -- Top 25 Words in the Movie Review Corpus",
fontsize = 30)
plt.xlabel("Words", fontsize = 30)
plt.ylabel("Frequency", fontsize = 30)
all_plot = sns.barplot(x = all_fdist.index, y = all_fdist.values,
ax=ax)
plt.xticks(rotation=60)
plt.show()
```

As seen in the preceding code, we start by collecting the words in the `movie_review` corpus by using the `words()` method of the corpus object. We then count the words with NLTK's `FreqDist()` function. At the same time, we are lowercasing the words and ignoring non-alpha words such as numbers and punctuation. Then, for clarity in the visualization, we'll restrict the words we'll look at to the most frequent 25 words. You may be interested in trying different values of `top_words` in the code block to see how the graph looks with more and fewer words.

When we call `plt.show()`, the distribution of word frequencies is displayed. This can be seen in *Figure 4.5*:

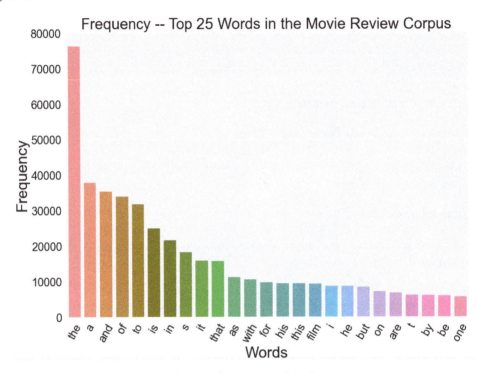

Figure 4.5 – Visualizing the most frequent words in the movie review corpus

As *Figure 4.5* shows, the most frequent word, not surprisingly, is **the**, which is about twice as frequent as the second most common word, **a**.

An alternative visualization of word frequencies that can also be helpful is a **word cloud**, where more frequent words are shown in a larger font. The following snippet shows the code for computing a word cloud from our word frequency distribution, `all_fdist`, and displaying it with Matplotlib:

```
from wordcloud import WordCloud
wordcloud = WordCloud(background_color = 'white',
```

```
                    max_words = 25,
                    relative_scaling = 0,
                    width = 600,height = 300,
                    max_font_size = 150,
                    colormap = 'Dark2',
                    min_font_size = 10).generate_from_
frequencies(all_fdist)

# Display the generated image:
plt.imshow(wordcloud, interpolation='bilinear')
plt.axis("off")
plt.show()
```

The resulting word cloud is shown in *Figure 4.6*. We can see that very frequent words such as **the** and **a** appear in very large fonts in comparison to other words:

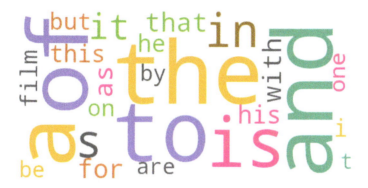

Figure 4.6 – A word cloud for the top 25 words in the movie review corpus

Notice that nearly all of the frequent words are words generally used in most English texts. The exception is **film**, which is to be expected in a corpus of movie reviews. Since most of these frequent words occur in the majority of texts, their occurrence won't enable us to distinguish different categories of texts. If we're dealing with a classification problem such as SA, we should consider removing these common words from texts before we try to train an SA classifier on this corpus. These kinds of words are called **stopwords**, and their removal is a common preprocessing step. We will discuss stopword removal in detail in *Chapter 5*.

POS frequencies

We can also look at the most frequent parts of speech with the following code. To reduce the complexity of the graph, we'll restrict the display to the 18 most common parts of speech. After we tag the words, we loop through the sentences, counting up the occurrences of each tag. Then, the list of tags is sorted with the most frequent tags first.

The parts of speech used in the NLTK POS tagging code are the widely used Penn Treebank parts of speech, documented at `https://www.cs.upc.edu/~nlp/SVMTool/PennTreebank.html`. This tagset includes 36 tags overall. Previous work on NLP has found that the traditional English parts of speech (noun, verb, adjective, adverb, conjunction, interjection, pronoun, and preposition) are not fine-grained enough for computational purposes, so additional parts of speech are normally added. For example, different forms of verbs, such as *walk*, *walks*, *walked*, and *walking*, are usually assigned different parts of speech. For example, *walk* is assigned the *VB*—or *Verb base* form—POS, but *walks* is assigned the *VBZ*—or *Verb, third - person singular present*—POS. Traditionally, these would all be called *verbs*.

First, we'll extract the sentences from the corpus, and then tag each word with its part of speech. It's important to perform POS tagging on an entire sentence rather than just individual words, because many words have multiple parts of speech, and the POS assigned to a word depends on the other words in its sentence. For example, *book* at the beginning of *book a flight* can be recognized and tagged as a verb, while in *I read the book*, *book* can be tagged as a noun. In the following code, we display the frequencies of the parts of speech in this corpus:

```
movie_reviews_sentences = movie_reviews.sents()
tagged_sentences = nltk.pos_tag_sents(movie_reviews_sentences)
total_counts = {}
for sentence in tagged_sentences:
    counts = Counter(tag for word,tag in sentence)
    total_counts = Counter(total_counts) + Counter(counts)
sorted_tag_list = sorted(total_counts.items(), key = lambda x:
x[1],reverse = True)
all_tags = pd.DataFrame(sorted_tag_list)
most_common_tags = all_tags.head(18)
# Setting figure and ax into variables
fig, ax = plt.subplots(figsize=(15,15))
all_plot = sns.barplot(x = most_common_tags[0], y = most_common_
tags[1], ax = ax)
plt.xticks(rotation = 70)
plt.title("Part of Speech Frequency  in Movie Review Corpus", fontsize
= 30)
plt.xlabel("Part of Speech", fontsize = 30)
plt.ylabel("Frequency", fontsize = 30)
plt.show()
```

In the movie review corpus, we can see that by far the most common tag is *NN* or *common noun*, followed by *IN* or *preposition or coordinating conjunction*, and *DT* or *determiner*.

The result, again using the Matplotlib and Seaborn libraries, is shown graphically in *Figure 4.7*:

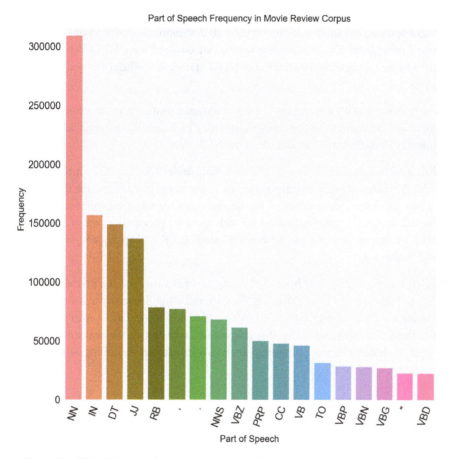

Figure 4.7 – Visualizing the most frequent parts of speech in the movie review corpus

We can look at other text properties such as the distribution of the lengths of the texts, and we can compare the properties of the positive and negative reviews to see if we can find some properties that distinguish the two categories. Do positive and negative reviews have different average lengths or different distributions of parts of speech? If we notice some differences, then we can make use of them in classifying new reviews.

Summary

In this chapter, we covered the major development tools and Python libraries that are used in NLP application development. We discussed the JupyterLab development environment and the GitHub software repository system. The major libraries that we covered were NLTK, spaCy, and Keras. Although this is by no means an exhaustive list of NLP libraries, it's sufficient to get a start on almost any NLP project.

We covered installation and basic usage for the major libraries, and we provided some suggested tips on selecting libraries. We summarized some useful auxiliary packages, and we concluded with a simple example of how the libraries can be used to do some NLP tasks.

The topics discussed in this chapter have given you a basic understanding of the most useful Python packages for NLP, which you will be using for the rest of the book. In addition, the discussion in this chapter has given you a start on understanding the principles for selecting tools for future projects. We have achieved our goal of getting you set up with tools for processing natural language, along with an illustration of some simple text processing using NLTK and spaCy and visualization with Matplotlib and Seaborn.

In the next chapter, we will look at how to identify and prepare data for processing with NLP techniques. We will discuss data from databases, the web, and other documents, as well as privacy and ethics considerations. For readers who don't have access to their own data or who wish to compare their results to those of other researchers, this chapter will also discuss generally available corpora. It will then go on to discuss preprocessing steps such as tokenization, stemming, stopword removal, and lemmatization.

5

Natural Language Data – Finding and Preparing Data

This chapter will teach you how to identify and prepare data for processing with natural language understanding techniques. It will discuss data from databases, the web, and different kinds of documents, as well as privacy and ethics considerations. The Wizard of Oz technique will be covered briefly. If you don't have access to your own data, or if you wish to compare your results to those of other researchers, this chapter will also discuss generally available and frequently used corpora. It will then go on to discuss preprocessing steps such as stemming and lemmatization.

This chapter will cover the following topics:

- Sources of data and annotation
- Ensuring privacy and observing ethical considerations
- Generally available corpora
- Preprocessing data
- Application-specific types of preprocessing
- Choosing among preprocessing techniques

Finding sources of data and annotating it

Data is where all **natural language processing (NLP)** projects start. Data can be in the form of written texts or transcribed speech. The purpose of data is to teach an NLP system what it should do when it's given similar data in the future. Specific collections of data are also called *corpora* or *datasets*, and we will often use these terms interchangeably. Very recently, large pretrained models have been developed that greatly reduce the need for data in many applications. However, these pretrained models, which will be discussed in detail in *Chapter 11*, do not in most cases eliminate the need for application-specific data.

Written language data can be of any length, ranging from very short texts such as tweets to multi-page documents or even books. Written language can be interactive, such as a record of a chatbot session between a user and a system, or it can be non-interactive, such as a newspaper article or blog. Similarly, spoken language can be long or short. Like written language, it can be interactive, such as a transcription of a conversation between two people, or non-interactive, such as broadcast news. What all NLP data has in common is that it is language, and it consists of words, in one or more human languages, that are used by people to communicate with each other. The goal of every NLP project is to take that data, process it by using specific algorithms, and gain information about what the author or authors of the data had in mind when they created the data.

One of the first steps in any NLP project is finding the right data. For that, you'll want to consider your goals in doing the project.

Finding data for your own application

If you have a specific, practical application in mind that you'd like to build, it's often easy to know what kind of data you need. For example, to build an enterprise assistant (one of the types of interactive applications shown in *Figure 1.3*), you'll need examples of conversations between users and either human agents or interactive systems of the kind you would like to build. These examples could already be available in the form of existing call center records.

The examples we will consider in the following subsections are as follows:

- Conversations in call centers
- Chat logs
- Databases
- Message boards and customer reviews

Let's begin!

Conversations in call centers

If you're planning to build a voice assistant that performs customer support, in many cases, the goal is to offload some work from an existing call center. In that case, there will frequently be many transcripts of previous calls between human agents and customers that can serve as data for training an application. The customers' questions will be examples of what the application will need to understand, and the agents' responses will be examples of how the system should respond. There will need to be an annotation process that assigns an overall intent, or customer goal, to each customer utterance. Most of the time, the annotation process will also need to label entities within the utterance. Before annotation of the intents and entities begins, there should be an initial design step where the intents and entities are determined.

Once the data has been annotated, it can be used as training data for applications. The training process will vary significantly depending on the technologies used. We will cover the use of the data in training in detail in *Chapter 9* and *Chapter 10*.

Chat logs

If you have a website that includes a chat window, the questions typed by customers can serve as training data, just like transcripts from call center conversations. The only difference in the data is that call center data is based on speech rather than typing. Otherwise, the annotation, design, and training processes will be very similar. The data itself will be a bit different since typed inputs tend to be shorter than spoken inputs and will contain spelling errors.

Databases

Databases can often be a good source of enterprise data. Databases often contain free text fields, where a user can enter any information they like. Free text fields are used for information such as narrative summaries of incident reports, and they often contain rich information that's not captured elsewhere in the other database fields. NLP can be extremely valuable for learning from this rich information because the contents of free text fields can be classified and analyzed using NLP techniques. This analysis can provide additional insight into the topic of the database.

Message boards and customer reviews

Like free text fields, message boards and customer support forums contain unformatted inputs from customers. Customer support message boards and customer product reviews can be valuable sources of data about product failures as well as customer attitudes about products. Although this information can be reviewed by human analysts, human analysis is time-consuming and expensive. In many cases, this information is abundant and can be used as the basis of very useful NLP applications.

So far, we've talked about application-specific data and how you can find and analyze it for the purposes of the application. On the other hand, sometimes you will be interested in analyzing data as part of a research project. The next section discusses where you can obtain data for a research project.

Finding data for a research project

If your goal is to contribute to the science of NLP, or if you just want to be able to compare your algorithms to other researchers' work, the kind of data you need will be quite different from the data discussed previously. Rather than finding data (possibly proprietary data inside your enterprise) that no one's ever used before, you will want to use data that's freely available to other researchers. Ideally, this data will already be public, but if not, it's important for you to make it available to others in the NLP research community so that they can replicate your work. Scientific conferences or journals will nearly always require that any newly collected data be made available before a paper can be presented or published. We will now discuss several ways of collecting new data.

Collecting data

Although there are many sources of pre-existing data, sometimes you will not find exactly what you need from currently available resources. Perhaps you want to work on a very specific technical topic, or perhaps you're interested in a rapidly changing topic such as COVID-19. You may need data that's specific to a particular local area or seasonal data that's only applicable to certain times of the year. For all of these reasons, it may be necessary to collect data specifically for your project.

There are several good ways to collect data under these circumstances, including **application programming interfaces** (**APIs**), crowdsourcing data, and Wizard of Oz. We will review these in the following sections.

APIs

Some social media services have feeds that can be accessed from APIs. Twitter, for example, has an API that developers can use to access Twitter services (`https://developer.twitter.com/en/docs/twitter-api`).

Crowdsourcing data

Some data can be generated by human workers using platforms such as Amazon's Mechanical Turk (`https://www.mturk.com/`). The data to be generated has to be clearly described for crowdworkers, along with any parameters or constraints on the data to be generated. Data that is easy for the average person to understand, as opposed to technical or specialized scientific data, is especially suitable for crowdsourcing. Crowdsourcing can be an effective way to obtain data; however, for this data to be useful, some precautions have to be taken:

- It's important to make sure that the crowdworkers have adequate instructions so that the data they create is sufficiently similar to the real data that the system will encounter during deployment. Crowdworker data has to be monitored to make sure that the crowdworkers are properly following instructions.

- Crowdworkers have to have the right knowledge in order to create data that's appropriate for specialized applications. For example, if the crowdworkers are to generate medical reports, they have to have medical backgrounds.

- Crowdworkers have to have sufficient knowledge of the language that they're generating data for in order to ensure that the data is representative of the real data. It is not necessary for the data to be perfectly grammatical – language encountered during deployment is not, especially speech data. Insisting on perfectly grammatical data can lead to stilted, unrealistic data.

Wizard of Oz

The **Wizard of Oz** (**WoZ**) method of collecting data is based on setting up a computer-human interaction situation where a system appears to be processing the user's inputs, but in fact, the processing is done by a human behind the scenes. The *Wizard of Oz* reference is from the line in the movie where the wizard says, "*Pay no attention to the man behind the curtain*," who is actually controlling a projection of what is supposed to be the wizard. The idea behind the technique is that if the user believes that a system is doing the processing, the user's behavior will represent how they would behave with an actual system. While the WoZ method can provide very high-quality data, it is expensive, since the setup must be carefully arranged so that subjects in the experiment are unaware that they're interacting with an automated system. You can find details about the WoZ paradigm at `https://en.wikipedia.org/wiki/Wizard_of_Oz_experiment`, and more information about conducting a WoZ experiment here: `https://www.answerlab.com/insights/wizard-of-oz-testing`.

In data collection, we are not only interested in the language itself but also in additional information that describes the data, or *metadata*. The next section will describe metadata in general, and then continue discussing an extremely important type of metadata, *annotation*.

Metadata

Datasets often include metadata. Metadata refers to information about the data rather than the data itself. Almost any information that data providers think might be useful in further processing can be included as metadata. Some of the most common types of metadata include the human language of the data, the speaker and time and place of speech for spoken data, and the author of written data. If spoken data is the result of a speech recognition process, the speech recognizer's confidence is usually included as metadata. The next section covers annotation, which is probably the most important type of metadata.

Annotation

One of the most important types of metadata is the intended NLP result for a text, or the annotation.

For the most part, newly collected data will need to be annotated, unless the experiment to be done involves unsupervised learning (more on unsupervised learning in *Chapter 12*). Annotation is the process of associating an input with the NLP result that the trained system is intended to produce. The system *learns* how to analyze data by processing the annotated examples, and then applies that learning to new, unannotated examples.

Since annotation is actually the *supervision* in supervised learning, data used in unsupervised learning experiments does not need to be annotated with the intended NLP result.

There are several software tools that can be used to annotate NLP text data. For example, the **General Architecture for Text Engineering** (**GATE**) (`https://gate.ac.uk/`) includes a well-tested user interface that enables annotators to assign meanings to documents and parts of documents.

In the following sections, we will "learn about the" transcription of speech data and the question of inter-annotator agreement.

Transcription

Transcription, or converting the speech contained in audio files to its written form, is a necessary annotation step for speech data. If the speech does not contain significant amounts of noise, commercial **automatic speech recognition** (**ASR**) systems such as Nuance Dragon (`https://www.nuance.com/dragon/business-solutions/dragon-professional-individual.html`) can provide a fairly accurate first pass at transcribing audio files. If you do use commercial ASR for transcription, the results should still be reviewed by a researcher in order to catch and correct any errors made by the ASR system. On the other hand, if the speech is very noisy or quiet, or if it contains speech from several people talking over each other, commercial ASR systems will probably make too many errors for the automatic transcription results to be useful. In that situation, manual transcription software such as TranscriberAG (`http://transag.sourceforge.net/`) can be used. You should keep in mind that manual transcription of noisy or otherwise problematic speech is likely to be a slow process because the transcriber has to first understand the speech in order to transcribe it.

Inter-annotator agreement

Annotators will not always agree on the correct annotation for any given data item. There can be general differences in opinion about the correct annotation, especially if the annotation instructions are unclear. In addition, annotators might not be thinking carefully about what the annotations should be, or the data that the annotators are being asked to annotate might be inherently subjective. For these reasons, annotation of the same data is often done by several annotators, especially if the data is subjective. Annotating an emotion associated with text or speech is a good example of data that has a lot of potential for disagreement. The degree of agreement among annotators is called **inter-annotator agreement** and is measured by what is known as the **kappa statistic**. The kappa statistic is preferable to just computing a percentage agreement because it takes into account the possibility that the annotators might agree by chance.

Natural language toolkit (**NLTK**) includes a package, `nltk.metrics.agreement`, that can be used to calculate inter-annotator agreement.

So far, we have discussed the process of obtaining your own data, but there are also many pre-existing datasets available that have already been annotated and are freely available. We'll discuss pre-existing data in the next section. When we start working directly with data in later chapters, we will be using pre-existing datasets.

Generally available corpora

The easiest way to get data is to use pre-existing corpora that cover the kind of problem that you're trying to solve. With pre-existing corpora, you will not need to collect the data, and you will probably not need to annotate it unless the pre-existing annotations are not suitable for your problem. Any privacy questions will have been addressed before the dataset was published. An added advantage to using pre-existing corpora is that other researchers have probably published papers describing their work on the corpus, which you can compare to your own work.

Fortunately, there are many standard preexisting datasets that you can download and work with, covering almost any NLP problem. Some are free and others are available for a fee.

Preexisting datasets are available from a number of organizations, such as the following:

- Linguistic Data Consortium (`https://www.ldc.upenn.edu/`): Provides a wide variety of text and speech data in many languages and also manages donated data.

- Hugging Face (`https://huggingface.co/`): Provides datasets in many languages, as well as NLP models. Some of the popular datasets available from Hugging Face include movie reviews, product reviews, and Twitter emotion categories.

- Kaggle (`https://www.kaggle.com/`): Provides many datasets, including user-contributed datasets.

- **European language resources association (ELRA)** (`http://www.elra.info/en/`): A European organization that provides multilingual data.

- Data distributed with NLP libraries such as NLTK and spaCy: NLP libraries include dozens of corpora of all sizes and languages. Much of the data is annotated in support of many different types of applications.

- Government data: Governments collect vast amounts of data, including text data, which is often publicly available and can be used for research.

- Librispeech `https://www.openslr.org/12`: A large dataset of read speech, based on audiobooks read for people with visual impairments. Primarily used for speech projects.

Up to this point, we have covered the topics of obtaining data and adding metadata, including annotation. Before data is used in NLP applications, we also need to ensure that it is used ethically. Ethical considerations are the topic of the next section.

Ensuring privacy and observing ethical considerations

Language data, especially data internal to an enterprise, may contain sensitive information. Examples that come to mind right away are medical and financial data. When an application deals with these kinds of topics, it is very likely to contain sensitive information about health or finances. Information can become even more sensitive if it is associated with a specific person. This is called **personally identifiable information (PII)**, which is defined by the United States Department of Labor as follows:

"*Any representation of information that permits the identity of an individual to whom the information applies to be reasonably inferred by either direct or indirect means*" (`https://www.dol.gov/general/ppii`). This is a broad and complex issue, a full treatment of which is out of the scope of this book. However, it's worth discussing a few important points specific to NLP applications that should be considered if you need to deal with any kind of sensitive data.

Ensuring the privacy of training data

In the case of generally available corpora, the data has typically been prepared so that sensitive information has been removed. If you are dealing with data you have obtained yourself, on the other hand, you will have to consider how to deal with sensitive information that you may encounter in the raw data.

One common strategy is to replace sensitive data with placeholders, such as <NAME>, <LOCATION>, and <PHONENUMBER>. This allows the training process to learn how to process the natural language without exposing any sensitive data. This should not affect the ability of the trained system to process natural language because it would be rare that an application would classify utterances differently depending on specific names or locations. If the classification depends on more specific information than just, for example, <LOCATION>, a more specific placeholder can be used, such as a city or country name. It is also actually helpful to use placeholders in training because it reduces the chances of overfitting the trained model on the specific names in the training data.

Ensuring the privacy of runtime data

At runtime, when an application is deployed, incoming data will naturally contain sensitive data that the users enter in order to accomplish their goals. Any precautions that are normally taken to secure data entered with forms (non-NLP data such as social security numbers or credit card numbers) will of course apply to data entered with natural language text and speech. In some cases, the handling of this data will be subject to regulations and legislation, which you will need to be aware of and follow.

Treating human subjects ethically

Natural language and speech data can be collected in the course of an experiment, such as a WoZ study, when human users or subjects provide speech and text for research. Universities and other research institutions have committees that review the planned procedures for any experiments with human subjects to ensure that the subjects are treated ethically. For example, subjects must provide informed consent to the experiment, they must not be harmed, their anonymity must be protected, deceptive practices must be avoided, and they must be able to withdraw from the study at any time. If you are collecting data in an experimental context, make sure to find out about the rules at your institution regarding approval by your human subjects committee or the equivalent. You may need to allocate extra lead time for the approval process to go through.

Treating crowdworkers ethically

If crowdworkers such as Amazon Mechanical Turk workers are used to create or annotate data, it is important to remember to treat them fairly – most importantly, to pay them fairly and on time – but also to make sure they have the right tools to do their jobs and to listen respectfully to any concerns or questions that they might have about their tasks.

Now that we have identified a source of data and discussed ethical issues, let's move on to what we can do to get the data ready to be used in an application, or preprocessing. We will first cover general topics in preprocessing in the next section, and then discuss preprocessing techniques for specific applications.

Preprocessing data

Once the data is available, it usually needs to be cleaned or preprocessed before the actual natural language processing begins.

There are two major goals in preprocessing data. The first goal is to remove items that can't be processed by the system – these might include items such as emojis, HTML markup, spelling errors, foreign words, or some Unicode characters such as *smart quotes*. There are a number of existing Python libraries that can help with this, and we'll be showing how to use them in the next section, *Removing non-text*. The second goal is addressed in the section called *Regularizing text*. We regularize text so that differences among words in the text that are not relevant to the application's goal can be ignored. For example, in some applications, we might want to ignore the differences between uppercase and lowercase.

There are many possible preprocessing tasks that can be helpful in preparing natural language data. Some are almost universally done, such as tokenization, while others are only done in particular types of applications. We will be discussing both types of preprocessing tasks in this section.

Different applications will need different kinds of preprocessing. Other than the most common preprocessing tasks, such as tokenization, exactly what kinds of preprocessing need to be done has to be carefully considered. A useful preprocessing step for one kind of application can completely remove essential information that's needed in another kind of application. Consequently, for each preprocessing step, it's important to think through its purpose and how it will contribute to the effectiveness of the application. In particular, you should use preprocessing steps thoughtfully if you will be using **large language models** (**LLMs**) in your NLP application. Since they are trained on normal (unregularized) text, regularizing input text to LLMs will make the input text less similar to the training text. This usually causes problems with machine learning models such as LLMs.

Removing non-text

Many natural language components are only able to process textual characters, but documents can also contain characters that are not text. Depending on the purpose of the application, you can either remove them from the text completely or replace them with equivalent characters that the system is able to process.

Note that the idea of *non-text* is not an all-or-none concept. What is considered *non-text* is to some extent application-specific. While the standard characters in the ASCII character set (https://www.ascii-code.com/) are clear examples of text, other characters can sometimes be considered non-text. For example, your application's *non-text* could include such items as currency symbols, math symbols, or texts written in scripts other than the main script of the rest of the text (such as a Chinese word in an otherwise English document).

In the next two sections, we will look at removing or replacing two common examples of non-text: emojis and smart quotes. These examples should provide a general framework for removing other types of non-text.

Removing emojis

One of the most common types of non-text is emojis. Social media posts are very likely to contain emojis, but they are a very interesting kind of text for NLP processing. If the natural language tools being used in your application don't support emojis, the emojis can either be removed or replaced with their text equivalents. In most applications, it is likely that you will want to remove or replace emojis, but it is also possible in some cases that your NLP application will be able to interpret them directly. In that case, you won't want to remove them.

One way to replace emojis is to use regular expressions that search for the Unicode (see `https://home.unicode.org/` for more information about Unicode) representations of emojis in the text. Still, the set of emojis is constantly expanding, so it is difficult to write a regular expression that covers all possibilities. Another approach is to use a Python library that directly accesses the Unicode data from `unicode.org`, which defines standard emojis (`https://home.unicode.org/`).

One package that can be used to remove or replace emojis is `demoji` (`https://pypi.org/project/demoji/`).

Using `demoji` is just a matter of installing it and running over text that may contain undesired emojis:

1. First, install `demoji`:

    ```
    $ pip install demoji
    ```

2. Then, given a text that contains emojis, run to replace the emojis with descriptions using the following code:

    ```
    demoji.replace_with_desc(text)
    ```

    Or, if you want to remove the emojis completely or replace them with a specific alternative of your choosing, you can run the following code:

    ```
    demoji.replace(text,replacement_text)
    ```

For example, *Figure 5.1* shows a text that includes an emoji of a birthday cake and shows how this can be replaced with the description `:birthday cake:`, or simply removed:

```
 1  import demoji
 2
 3  happy_birthday = "Happy birthday! 🎂"
 4
 5  text_with_emojis_replaced = demoji.replace_with_desc(happy_birthday)
 6  print(text_with_emojis_replaced)
 7
 8  text_with_emojis_removed = demoji.replace(happy_birthday,"")
 9  print(text_with_emojis_removed)
10
```

```
Happy birthday!:birthday cake:
Happy birthday!
```

Figure 5.1 – Replacing or removing emojis

Even if the emoji doesn't cause problems with running the software, leaving the emoji in place means that any meaning associated with the emoji will be ignored, because the NLP software will not understand the meaning of the emoji. If the emoji is replaced with a description, some of its meaning (for example, that the emoji represents a birthday cake) can be taken into account.

Removing smart quotes

Word processing programs sometimes automatically change typed quotation marks into *smart quotes*, or *curly quotes*, which look better than straight quotes, but which other software may not be prepared for. Smart quotes can cause problems with some NLP software that is not expecting smart quotes. If your text contains smart quotes, they can easily be replaced with the normal Python string replacement method, as in the following code:

```
text = "here is a string with "smart" quotes"
text = text.replace(""", "\"").replace(""","\"")
print(text)
here is a string with "smart" quotes
```

Note that the replacement straight quotes in the `replace` method need to be escaped with a backslash, like any use of literal quotes.

The previous two sections covered removing non-text items such as emojis and smart quotes. We will now talk about some techniques for regularizing text or modifying it to make it more uniform.

Regularizing text

In this section, we will be covering the most important techniques for regularizing text. We will talk about the goals of each technique and how to apply it in Python.

Tokenization

Nearly all NLP software operates on the level of words, so the text needs to be broken into words for processing to work. In many languages, the primary way of separating text into words is by whitespaces, but there are many special cases where this heuristic doesn't work. In *Figure 5.2*, you can see the code for splitting on whitespace and the code for tokenization using NLTK:

```python
import nltk
from nltk import word_tokenize

# a set of a few sentences to illustrate tokenization
text = ["Walk here.", "Walk  here.", "Don't walk here.", "$100"]

print("Split on white space")

for sentence in text:
    tokenized = sentence.split(" ")
    print(tokenized)

print("Using NLTK tokenization")

for sentence in text:
    tokenized = word_tokenize(sentence)
    print(tokenized)
```

Figure 5.2 – Python code for tokenization by splitting on whitespace and using NLTK's tokenization

Running the code in *Figure 5.2* results in the tokenizations shown in *Table 5.1*:

Example	Issue	Result from splitting on whitespace	NLTK result
`Walk here.`	Punctuation should not be included in the token	`['Walk', 'here.']`	`['Walk', 'here', '.']`
`Walk   here.`	Extra whitespace should not count as a token	`['Walk', '', 'here.']`	`['Walk', 'here', '.']`
`Don't walk here.`	The contraction "don't" should count as two tokens	`["Don't", 'walk', 'here.']`	`['Do', "n't", 'walk', 'here', '.']`
`$100`	The "$" should be a separate token	`['$100']`	`['$', '100']`

Table 5.1 – Tokenization results by splitting on whitespace and using NLTK's tokenization

Looking at *Table 5.1*, we can see that the simple heuristic of splitting the text on whitespace results in errors in some cases:

- In the first row of *Table 5.1*, we can compare the two approaches when punctuation occurs at the end of a token, but there is no whitespace between the token and punctuation. This means that just separating tokens on whitespace results in incorrectly including the punctuation in the token. This means that walk, walk,, walk., walk?, walk;, walk:, and walk! will all be considered to be different words. If all of those words appear to be different, any generalizations found in training based on one version of the word won't apply to the other versions.

- In the second row, we can see that if split on whitespace, two whitespaces in a row will result in an extra blank token. The result of this will be to throw off any algorithms that take into account the fact that two words are next to each other.

- Contractions also cause problems when splitting on whitespace is used for tokenization. Contractions of two words won't be separated into their components, which means that the **natural language understanding (NLU)** algorithms won't be able to take into account that *do not* and *don't* have the same meaning.

- Finally, when presented with words with monetary amounts or other measurements, the algorithms won't be able to take into account that *$100* and *100 dollars* have the same meaning.

It is tempting to try to write regular expressions to take care of these exceptions to the generalization that most words are surrounded by whitespace. However, in practice, it is very hard to capture all of the cases. As we attempt to cover more cases with regular expressions, the regular expressions will become more complex and difficult to maintain. For that reason, it is preferable to use a library such as NLTK's, which has been developed over many years and has been thoroughly tested. You can try different texts with this code to see what kinds of results you get with different tokenization approaches.

As *Table 5.1* shows, either way, the result will be a list of strings, which is a convenient form for further processing.

Lower casing

In languages that use uppercase and lowercase in their writing systems, most documents contain words with upper and lowercase letters. As in the case of tokenization, having the same word written in slightly different formats means that data from one format won't apply to data in the other formats. For example, *Walk*, *walk*, and *WALK* will all count as different words. In order to make them all count as the same word, the text is normally all in lowercase. This can be done by looping over a list of word tokens and applying the `lower()` Python function, as shown in *Figure 5.3*:

```
1  mixed_text = "WALK! Going for a walk is great exercise."
2  mixed_words = nltk.word_tokenize(mixed_text)
3  print(mixed_words)
4
5  lower_words = []
6  for mixed_word in mixed_words:
7      lower_words.append(mixed_word.lower())
8  print(lower_words)

['WALK', '!', 'Going', 'for', 'a', 'walk', 'is', 'great', 'exercise', '.']
['walk', '!', 'going', 'for', 'a', 'walk', 'is', 'great', 'exercise', '.']
```

Figure 5.3 – Converting text to all lowercase

Converting all words to lowercase has some drawbacks, however. Case differences are sometimes important for meaning. The biggest example of this is that it will make it hard for NLP software to tell the difference between proper names and ordinary words, which differ in case. This can cause errors in **part of speech (POS)** tagging or **named entity recognition (NER)** if the tagger or named entity recognizer is trained on data that includes case differences. Similarly, words can sometimes be written in all caps for emphasis. This may indicate something about the sentiment expressed in the sentence – perhaps the writer is excited or angry – and this information could be helpful in sentiment analysis. For these reasons, the position of each preprocessing step in a pipeline should be considered to make sure no information is removed before it's needed. This will be discussed in more detail in the *Text preprocessing pipeline* section later in this chapter.

Stemming

Words in many languages appear in different forms depending on how they're used in a sentence. For example, English nouns have different forms depending on whether they're singular or plural, and English verbs have different forms depending on their tense. English has only a few variations, but other languages sometimes have many more. For example, Spanish has dozens of verb forms that indicate past, present, or future tenses or whether the subject of the verb is first, second, or third person, or singular or plural. Technically, in linguistics, these different forms are referred to as **inflectional morphology**. The endings themselves are referred to as **inflectional morphemes**.

Of course, in speech or text that is directed toward another person, these different forms are very important in conveying the speaker's meaning. However, if the goal of our NLP application is to classify documents into different categories, paying attention to different forms of a word is likely not to be necessary. Just as with punctuation, different forms of words can cause them to be treated as completely separate words by NLU processors, despite the fact that the words are very similar in meaning.

Stemming and **lemmatization** are two similar methods of regularizing these different forms. Stemming is the simpler approach, so we will look at that first. Stemming basically means removing specific letters that end some words and that are frequently, but not always, inflectional morphemes – for example, the *s* at the end of *walks*, or the *ed* at the end of *walked*. Stemming algorithms don't have any knowledge of the actual words in a language; they're just guessing what may or may not be an ending. For that reason, they make a lot of mistakes. They can make mistakes by either removing too many letters or not enough letters, resulting in two words being collapsed that are actually different words, or not collapsing words that should be treated as the same word.

`PorterStemmer` is a widely used stemming tool and is built into NLTK. It can be used as shown in *Figure 5.4*:

```
import nltk
from nltk.stem.porter import PorterStemmer
stemmer = PorterStemmer()
text_to_stem = "Going for a walk is the best exercise. I've walked every evening this week."
tokenized_to_stem = nltk.word_tokenize(text_to_stem)
stemmed = [stemmer.stem(w) for w in tokenized_to_stem]
print(stemmed)

['go', 'for', 'a', 'walk', 'is', 'the', 'best', 'exercis', '.', 'I', "'ve", 'walk', 'everi',
'even', 'thi', 'week', '.']
```

Figure 5.4 – Results from stemming with PorterStemmer

Note that the results include a number of mistakes. While `walked` becomes `walk`, and `going` becomes `go`, which are good results, the other changes that the stemmer made are errors:

- `exercise` → `exercis`
- `every` → `everi`
- `evening` → `even`
- `this` → `thi`

The Porter stemmer also only works for English because its stemming algorithm includes heuristics, such as removing the *s* at the end of words, that only apply to English. NLTK also includes a multilingual stemmer, called the Snowball stemmer, that can be used with more languages, including Arabic, Danish, Dutch, English, Finnish, French, German, Hungarian, Italian, Norwegian, Portuguese, Romanian, Russian, Spanish, and Swedish.

However, since these stemmers don't have any specific knowledge of the words of the languages they're applied to, they can make mistakes, as we have seen. A similar but more accurate approach actually makes use of a dictionary, so that it doesn't make errors like the ones listed previously. This approach is called **lemmatization**.

Lemmatizing and part of speech tagging

Lemmatization, like stemming, has the goal of reducing variation in the words that occur in the text. However, lemmatization actually replaces each word with its root word (found by looking the word up in a computational dictionary) rather than simply removing what looks like suffixes. However, identifying the root word often depends on the part of speech, and lemmatization can be inaccurate if it doesn't know the word's part of speech. The part of speech can be identified through **part of speech tagging**, covered in *Chapter 3*, which assigns the most probable part of speech to each word in a text. For that reason, lemmatization and part of speech tagging are often performed together.

For the dictionary in this example, we'll use WordNet, developed at Princeton University (`https://wordnet.princeton.edu/`), an important source of information about words and their parts of speech. The original WordNet was developed for English, but WordNets for other languages have also been developed. We briefly mentioned WordNet in the *Semantic analysis* section of *Chapter 3*, because WordNet contains semantic information as well as information about parts of speech.

In this example, we'll just use the part of speech information, not the semantic information. *Figure 5.5* shows importing the WordNet lemmatizer, the tokenizer, and the part of speech tagger. We then have to align the names of the parts of speech between WordNet and the part of speech tagger, because the part of speech tagger and WordNet don't use the same names for the parts of speech. We then go through the text, lemmatizing each word:

```
 1  import nltk
 2  nltk.download("wordnet")
 3  from nltk.stem.wordnet import WordNetLemmatizer
 4  from nltk import word_tokenize, pos_tag
 5  from nltk.corpus import wordnet
 6  from collections import defaultdict
 7
 8  # align names for parts of speech between WordNet and part of speech tagger.
 9  tag_map = defaultdict(lambda: wordnet.NOUN)
10  tag_map["J"] = wordnet.ADJ
11  tag_map["V"] = wordnet.VERB
12  tag_map["R"] = wordnet.ADJ
13
14  lemmatizer = WordNetLemmatizer()
15  text_to_lemmatize = (
16      "going for a walk is the best exercise. i've walked every evening this week"
17  )
18  print("text to lemmatize is: \n", text_to_lemmatize)
19
20  tokens_to_lemmatize = nltk.word_tokenize(text_to_lemmatize)
21  lemmatized_result = ""
22  for token, tag in pos_tag(tokens_to_lemmatize):
23      lemma = lemmatizer.lemmatize(token, tag_map[tag[0]])
24      lemmatized_result = lemmatized_result + " " + lemma
25  print("lemmatized result is: \n", lemmatized_result)
```

```
[nltk_data] Downloading package wordnet to
[nltk_data]     C:\Users\dahl\AppData\Roaming\nltk_data...
[nltk_data]   Package wordnet is already up-to-date!
text to lemmatize is:
 going for a walk is the best exercise. i've walked every evening this week
lemmatized result is:
  go for a walk be the best exercise . i 've walk every evening this week
```

Figure 5.5 – Lemmatization for "going for a walk is the best exercise. i've walked every evening this week"

As the lemmatized result shows, many of the words in the input text have been replaced by their lemmas – going is replaced with go, is is replaced with be, and walked is replaced with walk.

Note that evening hasn't been replaced by *even*, as it was in the stemming example. If *evening* had been the present participle of the verb *even*, it would have been replaced by even, but *even* isn't the root word for *evening* here. In this case, *evening* just refers to the time of day.

Stopword removal

Stopwords are extremely common words that are not helpful in distinguishing documents and so they are often removed in classification applications. However, if the application involves any kind of detailed sentence analysis, these common words are needed so that the system can figure out what the analysis should be.

Normally, stopwords include words such as pronouns, prepositions, articles, and conjunctions. Which words should be considered to be stopwords for a particular language is a matter of judgment. For example, spaCy has many more English stopwords (326) than NLTK (179). These specific stopwords were chosen by the spaCy and NLTK developers because they thought those stopwords would be useful in practice. You can use whichever one you find more convenient.

Let's take a look at the stopwords provided by each system.

First, to run the NLTK and spaCy systems, you may need to do some preliminary setup. If you are working in a command-line or terminal environment, you can ensure that NLTK and spaCy are available by entering the following commands on the command line:

1. `pip install -U pip setuptools wheel`

2. `pip install -U spacy`

3. `python -m spacy download en_core_web_sm`

On the other hand, if you're working in the (recommended) Jupyter Notebook environment covered in *Chapter 4*, you can enter the same commands in a Jupyter code cell but precede each command with `!`.

Once you've confirmed that your NLTK and spaCy environments are set up, you can look at the NLTK stopwords by running the code in *Figure 5.6*:

```
1 from nltk.corpus import stopwords
2 nltk.download('stopwords')
3 nltk_stopwords = nltk.corpus.stopwords.words('english')
4 print(nltk_stopwords)

['i', 'me', 'my', 'myself', 'we', 'our', 'ours', 'ourselves', 'you', "you're", "you've", "you'll", "you'd",
es', 'he', 'him', 'his', 'himself', 'she', "she's", 'her', 'hers', 'herself', 'it', "it's", 'its', 'itself'
hemselves', 'what', 'which', 'who', 'whom', 'this', 'that', "that'll", 'these', 'those', 'am', 'is', 'are',
'have', 'has', 'had', 'having', 'do', 'does', 'did', 'doing', 'a', 'an', 'the', 'and', 'but', 'if', 'or', '
'at', 'by', 'for', 'with', 'about', 'against', 'between', 'into', 'through', 'during', 'before', 'after', '
```

Figure 5.6 – Viewing the first few stopwords for NLTK

Note that *Figure 5.6* just shows the first few stopwords. You can see them all by running the code in your environment.

To see the spaCy stopwords, run the code in *Figure 5.7*, which also just shows the first few stopwords:

```
1 import spacy
2 nlp = spacy.load('en_core_web_sm')
3 spacy_stopwords = nlp.Defaults.stop_words
4 print(spacy_stopwords)
```

```
{''m', 'now', 'very', 'elsewhere', 'out', 'used', 'besides', 'mine', ''re', 'top', 'each', 'whence', ''d', 'enough'
'would', 'mostly', 'that', 'such', 'for', 'everything', 'some', 'n't', 'its', 'becomes', "'m", 'also', 'own', 'call
g', 'no', 'side', 'myself', 'toward', 'amount', 'did', 'otherwise', 'ca', 'n't', 'how', 'so', 'which', 'whatever',
s', 'almost', 'eleven', 'nevertheless', 'seemed', 'he', 'anyhow', 'front', 'always', 'therein', 'please', 'fifteen'
l', 'although', 'behind', 'bottom', 'much', 'yours', 'too', 'herein', 'made', 'without', 'after', 'hence', 'are', '
```

Figure 5.7 – Viewing the first few stopwords for spaCy

Comparing the stopwords provided by both packages, we can see that the two sets have a lot in common, but there are differences as well. In the end, the stopwords you use are up to you. Both sets work well in practice.

Removing punctuation

Removing punctuation can also be useful since punctuation, like stopwords, appears in most documents and therefore doesn't help distinguish document categories.

Punctuation can be removed by defining a string of punctuation symbols and removing items in that string with regular expressions, or by removing every non-alphanumeric word in the text. The latter approach is more robust because it's easy to overlook an uncommon punctuation symbol.

The code to remove punctuation can be seen in *Figure 5.8*:

```
# define a sample text and tokenize it
text_to_remove_punct = "going for a walk is the best exercise!! i've walked, i believe, every evening this week."
tokens_to_remove_punct = nltk.word_tokenize(text_to_remove_punct)

# remove punctuation
tokens_no_punct = [word for word in tokens_to_remove_punct if word.isalnum()]
print(tokens_no_punct)
```

```
['going', 'for', 'a', 'walk', 'is', 'the', 'best', 'exercise', 'i', 'walked', 'i', 'believe', 'every', 'evening',
'this', 'week']
```

Figure 5.8 – Removing punctuation

The original text is the value of the `text_to_remove_punct` variable, in *Figure 5.8*, which contains several punctuation marks – specifically, an exclamation mark, a comma, and a period. The result is the value of the `tokens_no_punct` variable, shown in the last line.

Spelling correction

Correcting misspelled words is another way of removing noise and regularizing text input. Misspelled words are much less likely to have occurred in training data than correctly spelled words, so they will be harder to recognize when a new text is being processed. In addition, any training that includes the correctly spelled version of the word will not recognize or be able to make use of data including the incorrectly spelled word. This means that it's worth considering adding a spelling correction preprocessing step to the NLP pipeline. However, we don't want to automatically apply spelling correction in every project. Some of the reasons not to use spelling correction are the following:

- Some types of text will naturally be full of spelling mistakes – for example, social media posts. Because spelling mistakes will occur in texts that need to be processed by the application, it might be a good idea not to try to correct spelling mistakes in either the training or runtime data.

- Spelling error correction doesn't always do the right thing, and a spelling correction that results in the wrong word won't be helpful. It will just introduce errors into the processing.

- Some types of texts include many proper names or foreign words that aren't known to the spell checker, which will try to correct them. Again, this will just introduce errors.

If you do choose to use spelling corrections, there are many spell checkers available in Python. One recommended spell checker is `pyspellchecker`, which can be installed as follows:

```
$ pip install pyspellchecker
```

The spell-checking code and the result can be seen in *Figure 5.9*:

```python
from spellchecker import SpellChecker

spell_checker = SpellChecker()

# find words that may represent spelling errors
text_to_spell_check = "Ms. Ramalingam voted agains the bill"
tokens_to_spell_check = nltk.word_tokenize(text_to_spell_check)
spelling_errors = spell_checker.unknown(tokens_to_spell_check)

for misspelled in spelling_errors:
    # Get the one `most likely` answer
    print(misspelled, " should be", spell_checker.correction(misspelled))


ms.   should be is
ramalingam   should be None
agains   should be against
```

Figure 5.9 – Spell-checking with pyspellchecker

You can see from *Figure 5.9* that it's easy to make mistakes with spell-checking. `pyspellchecker` correctly changes `agains` to `against`, but it also made a mistake correcting `Ms.` to `is` and it didn't know anything about the name `Ramalingam`, so it didn't have a correction for that.

Keep in mind that input generated by **ASR** will not contain spelling errors, because ASR can only output words in their dictionaries, which are all correctly spelled. Of course, ASR output can contain mistakes, but those mistakes will just be the result of substituting a wrong word for the word that was actually spoken, and they can't be corrected by spelling correction.

Also note that stemming and lemmatization can result in tokens that aren't real words, which you don't want to be corrected. If spelling correction is used in a pipeline, make sure that it occurs before stemming and lemmatization.

Expanding contractions

Another way to increase the uniformity of data is to expand contractions – that is, words such as *don't* would be expanded to their full form, *do not*. This will allow the system to recognize an occurrence of *do* and an occurrence of *not* when it finds *don't*.

So far, we have reviewed many generic preprocessing techniques. Next, let's move on to more specific techniques that are only applicable to certain types of applications.

Application-specific types of preprocessing

The preprocessing topics we have covered in the previous sections are generally applicable to many types of text in many applications. Additional preprocessing steps can also be used in specific applications, and we will cover these in the next sections.

Substituting class labels for words and numbers

Sometimes data includes specific words or tokens that have equivalent semantics. For example, a text corpus might include the names of US states, but for the purposes of the application, we only care that *some* state was mentioned – we don't care which one. In that case, we can substitute a *class token* for the specific state name. Consider the interaction in *Figure 5.10*:

System: where do you live?

User: I live in Texas

Class token substitution: I live in <state_name>.

Figure 5.10 – Class token substitution

If we substitute the class token, `<state_name>`, for `Texas`, all of the other state names will be easier to recognize, because instead of having to learn 50 states, the system will only have to learn about the general class, `<state_name>`.

Another reason to use class tokens is if the texts contain alphanumeric tokens such as dates, phone numbers, or social security numbers, especially if there are too many to enumerate. Class tokens such as `<social_security_number>` can be substituted for the actual numbers. This has the added benefit of masking, or *redacting*, sensitive information such as specific social security numbers.

Redaction

As we discussed in the *Ensuring privacy and observing ethical considerations* section, data can contain sensitive information such as people's names, health information, social security numbers, or telephone numbers. This information should be redacted before the data is used in training.

Domain-specific stopwords

Both NLTK and spaCy have the capability to add and remove stopwords from their lists. For example, if your application has some very common domain-specific words that aren't on the built-in stopword list, you can add these words to the application-specific stopword list. Conversely, if some words that are normally stopwords are actually meaningful in the application, these words can be removed from the stopwords.

A good example is the word *not*, which is a stopword for both NLTK and spaCy. In many document classification applications, it's fine to consider *not* as a stopword; however, in applications such as sentiment analysis, *not* and other related words (for example, *nothing* or *none*) can be important clues for a negative sentiment. Removing them can cause errors if sentences such as *I do not like this product* becomes *I do like this product*. In that case, you should remove *not* and other negative words from the list of stopwords you are using.

Remove HTML markup

If the application is based on web pages, they will contain HTML formatting tags that aren't useful in NLP. The Beautiful Soup library (`https://www.crummy.com/software/BeautifulSoup/bs4/doc/`) can perform the task of removing HTML tags. While Beautiful Soup has many functions for working with HTML documents, for our purposes, the most useful function is `get_text()`, which extracts the text from an HTML document.

Data imbalance

Text classification, where the task is to assign each document to one of a set of classes, is one of the most common types of NLP applications. In every real-life classification dataset, some of the classes will have more examples than others. This problem is called **data imbalance**. If the data is severely imbalanced, this will cause problems with machine learning algorithms. Two common techniques for addressing data imbalance are **oversampling** and **undersampling**. Oversampling means that some of the items in the less frequent classes are duplicated, and undersampling means that some of the items in the more common classes are removed. Both approaches can be used at the same time – frequent classes can be undersampled while infrequent classes can be oversampled. We will be discussing this topic in detail in *Chapter 14*.

Using text preprocessing pipelines

Getting your data ready for NLP often involves multiple steps, and each step takes the output of the previous step and adds new information, or in general, gets the data further prepared for NLP. A sequence of preprocessing steps like this is called a **pipeline**. For example, an NLP pipeline could include tokenization followed by lemmatization and then stopword removal. By adding and removing steps in a pipeline, you can easily experiment with different preprocessing steps and see whether they make a difference in the results.

Pipelines can be used to prepare both training data to be used in learning a model, as well as test data to be used at runtime or during testing. In general, if a preprocessing step is always going to be needed (for example, tokenization), it's worth considering using it on the training data once and then saving the resulting data as a dataset. This will save time if you're running experiments with different configurations of preprocessing steps to find the best configuration.

Choosing among preprocessing techniques

Table 5.2 is a summary of the preprocessing techniques described in this chapter, along with their advantages and disadvantages. It is important for every project to consider which techniques will lead to improved results:

Preprocessing Technique	Definition	Advantages	Disadvantages	Comments
Remove/replace non-text	Remove/replace emojis, HTML markup, and other non-text	Non-text can sometimes cause processing errors, but at best, introduces uninformative noise		Removal is unlikely to introduce errors
Correct data imbalance	Balance classes by undersampling or oversampling	Better performance of machine learning algorithms that expect balanced topics		Not necessary unless the imbalance is severe
Tokenization	Break text into individual tokens	Required for further processing	Overly simplistic tokenization techniques can lead to errors	Nearly all NLP algorithms operate on tokens
Spelling correction	Replace misspelled words with correct spellings	Make it possible to take advantage of more data	Inaccurate spelling correction can introduce errors	
Part of speech tagging	Assign each token its most probable part of speech	Improves lemmatization accuracy; helps separate words with different senses	Inaccurate part of speech tagging can introduce errors in downstream processing	
Stemming and lemmatization	Remove endings so that words with the same meaning look the same	Provides more data by grouping words with the same meaning but different endings	Inaccurate stemming and lemmatization can introduce errors	Must not be done before syntactic parsing as it will remove important information
Lowercase	Change all letters to their lowercase form	Provides more data by grouping together words that are the same but with different cases	Can cause errors in named entity recognition algorithms that use capitalization as an indication of proper names	
Remove punctuation	Remove periods, commas, question marks, and similar punctuation	Removes tokens that don't distinguish documents	Could cause errors in sentiment analysis if "?" and "!" are used to detect sentiments	
Remove stopwords	Remove extremely common words such as pronouns, prepositions and articles	Removes very common words that don't distinguish documents		Must not remove stopwords prior to part of speech tagging or syntactic parsing because it will remove important information
Expand contractions	Replace contractions such as "don't" with their uncontracted forms ("do not")	Provides more data by grouping together contracted with uncontracted forms		

Table 5.2 – Advantages and disadvantages of preprocessing techniques

Many techniques, such as spelling correction, have the potential to introduce errors because the technology is not perfect. This is particularly true for less well-studied languages, for which the relevant algorithms can be less mature than those of better-studied languages.

It is worth starting with an initial test with only the most necessary techniques (such as tokenization) and introducing additional techniques only if the results of the initial test are not good enough. Sometimes, the errors introduced by preprocessing can cause the overall results to get worse. It is important to keep evaluating results during the investigation to make sure that results aren't getting worse. Evaluation will be discussed in detail in *Chapter 13*.

Summary

In this chapter, we covered how to find and use natural language data, including finding data for a specific application as well as using generally available corpora.

We discussed a wide variety of techniques for preparing data for NLP, including annotation, which provides the foundation for supervised learning. We also discussed common preprocessing steps that remove noise and decrease variation in the data and allow machine learning algorithms to focus on the most informative differences among different categories of texts. Another important set of topics covered in this chapter had to do with privacy and ethics – how to ensure the privacy of information included in text data and how to ensure that crowdsourcing workers who are generating data or who are annotating data are treated fairly.

The next chapter will discuss exploratory techniques for getting an overall picture of a dataset, such as summary statistics (word frequencies, category frequencies, and so on). It will also discuss visualization tools (such as matplotlib) that can provide the kinds of insights that can be best obtained by looking at graphical representations of text data. Finally, it will discuss the kinds of decisions that can be made based on visualization and statistical results.

6

Exploring and Visualizing Data

Exploring and visualizing data are essential steps in the process of developing a **natural language understanding** (NLU) application. In this chapter, we will explore techniques for **data exploration**, such as visualizing word frequencies, and techniques for visualizing document similarity. We will also introduce several important visualization tools, such as Matplotlib, Seaborn, and WordCloud, that enable us to graphically represent data and identify patterns and relationships within our datasets. By combining these techniques, we can gain valuable perspectives into our data, make informed decisions about the next steps in our NLU processing, and ultimately, improve the accuracy and effectiveness of our analyses. Whether you're a data scientist or a developer, data exploration and visualization are essential skills for extracting actionable insights from text data in preparation for further NLU processing.

In this chapter, we will cover several topics related to the initial exploration of data, especially visual exploration, or **visualization**. We will start by reviewing a few reasons why getting a visual perspective of our data can be helpful. This will be followed by an introduction to a sample dataset of movie reviews that we will be using to illustrate our techniques. Then, we will look at techniques for data exploration, including summary statistics, visualizations of word frequencies in a dataset, and measuring document similarity. We will follow with a few general tips on developing visualizations and conclude with some ideas about using information from visualizations to make decisions about further processing.

The topics we will cover are as follows:

- Why visualize?
- Data exploration
- General considerations for developing visualizations
- Using information from visualization to make decisions about processing

Why visualize?

Visualizing data means displaying data in a graphical format such as a chart or graph. This is almost always a useful precursor to training a **natural language processing** (**NLP**) system to perform a specific task because it is typically very difficult to see patterns in large amounts of text data. It is often much easier to see overall patterns in data visually. These patterns might be very helpful in making decisions about the most applicable text-processing techniques.

Visualization can also be useful in understanding the results of NLP analysis and deciding what the next steps might be. Because looking at the results of NLP analysis is not an initial exploratory step, we will postpone this topic until *Chapter 13* and *Chapter 14*.

In order to explore visualization, in this chapter, we will be working with a dataset of text documents. The text documents will illustrate a binary classification problem, which will be described in the next section.

Text document dataset – Sentence Polarity Dataset

The Sentence Polarity Dataset is a commonly used dataset that consists of movie reviews originally from the **Internet Movie Database** (**IMDb**). The reviews have been categorized in terms of positive and negative reviews. The task that is most often used with these reviews is classifying the reviews into positive and negative. The data was collected by a team at Cornell University. More information about the dataset and the data itself can be found at https://www.cs.cornell.edu/people/pabo/movie-review-data/.

Natural Language Toolkit (**NLTK**) comes with this dataset as one of its built-in corpora. There are 1,000 positive reviews and 1,000 negative reviews.

Here's an example of a positive review from this dataset:

```
kolya is one of the richest films i've seen in some time .
zdenek sverak plays a confirmed old bachelor ( who's likely to remain
so ) , who finds his life as a czech cellist increasingly impacted by
the five-year old boy that he's taking care of .
though it ends rather abruptly-- and i'm whining , 'cause i wanted to
spend more time with these characters-- the acting , writing , and
production values are as high as , if not higher than , comparable
american dramas .
this father-and-son delight-- sverak also wrote the script , while his
son , jan , directed-- won a golden globe for best foreign language
film and , a couple days after i saw it , walked away an oscar .
in czech and russian , with english subtitles .
```

Here's an example of a negative review:

```
claire danes , giovanni ribisi , and omar epps make a likable trio of
protagonists , but they're just about the only palatable element of
the mod squad , a lame-brained big-screen version of the 70s tv show .
the story has all the originality of a block of wood ( well , it would
if you could decipher it ) , the characters are all blank slates ,
and scott silver's perfunctory action sequences are as cliched as they
come .
by sheer force of talent , the three actors wring marginal enjoyment
from the proceedings whenever they're on screen , but the mod squad is
just a second-rate action picture with a first-rate cast .
```

We can look at the thousands of examples in our dataset, but it will be hard to see any large-scale patterns by looking at individual examples; there is just too much data for us to separate the big picture from the details. The next sections will cover tools that we can use to identify these large-scale patterns.

Data exploration

Data exploration, which is sometimes also called **exploratory data analysis** (**EDA**), is the process of taking a first look at your data to see what kinds of patterns there are to get an overall perspective on the full dataset. These patterns and overall perspective will help us identify the most appropriate processing approaches. Because some NLU techniques are very computationally intensive, we want to ensure that we don't waste a lot of time applying a technique that is inappropriate for a particular dataset. Data exploration can help us narrow down the options for techniques at the very beginning of our project. Visualization is a great help in data exploration because it is a quick way to get the big picture of patterns in the data.

The most basic kind of information about a corpus that we would want to explore includes information such as the number of words, the number of distinct words, the average length of documents, and the number of documents in each category. We can start by looking at the frequency distributions of words. We will cover several different ways of visualizing word frequencies in our datasets and then look at some measurements of document similarity.

Frequency distributions

Frequency distributions show how many items of a specific type occur in some context. In our case, the context is a dataset, and we will be looking at the frequencies of *words* and then *ngrams* (sequences of words) that occur in our dataset and subsets of the dataset. We'll start by defining word frequency distributions and doing some preprocessing. We'll then visualize the word frequencies with Matplotlib tools and WordCloud, and then apply the same techniques to ngrams. Finally, we'll cover some techniques for visualizing document similarity: bag of words (**BoW**) and k-means clustering.

Word frequency distributions

The words and their frequencies that occur in a corpus are often very informative. It's a good idea to take a look at this information before getting too far into an NLP project. Let's take a look at computing this information using NLTK. We start by importing NLTK and the movie reviews dataset, as shown in *Figure 6.1*:

```
1  import nltk
```

```
1  #import the training data
2  from nltk.corpus import movie_reviews
```

Figure 6.1 – Importing NLTK and the movie reviews dataset

Figure 6.2 shows code that we can use to collect the words in the `movie_reviews` dataset into a list and take a look at some of the words in the following steps:

1. To collect the words in the corpus into a Python list, NLTK provides a helpful function, `words()`, which we use here.

2. Find out the total number of words in the dataset by using the normal Python `len()` function for finding the size of a list, with the list of words as a parameter.

3. Print the list of words:

```
3  corpus_words = movie_reviews.words()
4  print(len(corpus_words))
5  print(corpus_words)
```

Figure 6.2 – Making a list of the words in the corpus, counting them, and displaying the first few

The result of counting the words and printing the first few words is shown here:

```
1583820
['plot', ':', 'two', 'teen', 'couples', 'go', 'to', ...]
```

We won't see more words than the first few (`'plot'`, `':'`, `'two'`, and so on) because there are too many to show. By printing the length of the list, we can see that there are 1,583,820 individual words in the dataset, which is certainly too many to list. What we can also notice is that this list of words includes punctuation, specifically `:`. Since punctuation is so common, looking at it is unlikely to yield any insights into differences between documents. In addition, including punctuation will also make it harder to see the words that actually distinguish categories of documents. So, let's remove punctuation.

Figure 6.3 shows a way to remove punctuation with the following steps:

1. Initialize an empty list where we will store the new list of words after removing punctuation (words_ no_punct).

2. Iterate through the list of words (corpus_words).

3. Keep only the alphanumeric words. This code uses the Python string function, string. isalnum(), at line 4 to detect alphanumeric words, which are added to the words_no_ punct list that we previously initialized.

4. Once we remove the punctuation, we might be interested in the frequencies of occurrence of the non-punctuation words. Which are the most common words in our corpus? NLTK provides a useful FreqDist() function (frequency distribution) that computes word frequencies. This function is applied at line 6 in *Figure 6.3*.

5. We can then see the most common words in the frequency distribution using the NLTK most_common() method, which takes as a parameter how many words we want to see, shown in line 8.

6. Finally, we print the 50 most common words:

```
1  # remove punctuation
2  words_no_punct = []
3  for word in corpus_words:
4      if word.isalnum():
5          words_no_punct.append(word)
6  freq = nltk.FreqDist(words_no_punct)
7  #common words
8  print("Common Words:", freq.most_common(50))
```

```
Common Words: [('the', 76529), ('a', 38106), ('and', 35576), ('of', 34123), ('to', 31
937), ('is', 25195), ('in', 21822), ('s', 18513), ('it', 16107), ('that', 15924), ('a
s', 11378), ('with', 10792), ('for', 9961), ('his', 9587), ('this', 9578), ('film', 9
517), ('i', 8889), ('he', 8864), ('but', 8634), ('on', 7385), ('are', 6949), ('t', 64
10), ('by', 6261), ('be', 6174), ('one', 5852), ('movie', 5771), ('an', 5744), ('wh
o', 5692), ('not', 5577), ('you', 5316), ('from', 4999), ('at', 4986), ('was', 4940),
('have', 4901), ('they', 4825), ('has', 4719), ('her', 4522), ('all', 4373), ('ther
e', 3770), ('like', 3690), ('so', 3683), ('out', 3637), ('about', 3523), ('up', 340
5), ('more', 3347), ('what', 3322), ('when', 3258), ('which', 3161), ('or', 3148),
('she', 3141)]
```

Figure 6.3 – The most common 50 words in the movie review database and their frequencies

In *Figure 6.3*, we can easily see that the is the most common word, with 76,529 occurrences in the dataset, which is not surprising. However, it's harder to see the frequencies of the less common words. It's not easy to see, for example, the tenth most common word, and how much more common it is than the eleventh most common word. This is where we bring in our visualization tools.

The frequencies that we computed in *Figure 6.3* can be plotted with the `plot()` function for frequency distributions. This function takes two parameters:

1. The first parameter is the number of words we want to see – in this case, it is `50`.

2. The `cumulative` parameter determines whether we just want to see the frequency of each of our 50 words (`cumulative=False`). If we want to see the cumulative frequencies for all the words, we would set this parameter to `True`.

    To plot the frequency distribution, we can simply add a call to `plot()` as follows:

    ```
    freq.plot(50, cumulative = False)
    ```

The result is shown in *Figure 6.4*, a plot of the frequencies of each word, with the most frequent words first:

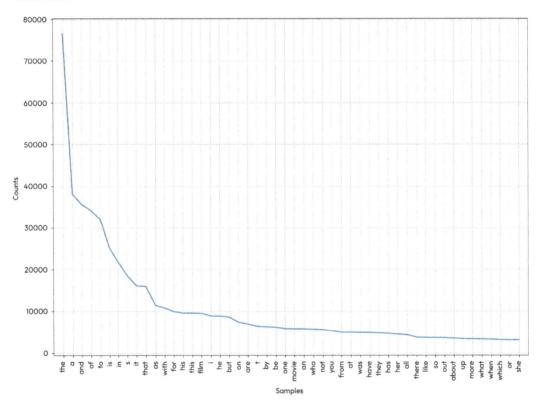

Figure 6.4 – Frequency distribution of words in the movie review corpus

Try `freq.plot()` with different numbers of words and with `cumulative = True`. What happens when you ask for more and more words? You will see that looking at rarer and rarer words (by increasing the value of the first argument to larger numbers) does not provide much insight.

While the graph in *Figure 6.4* gives us a much clearer idea of the frequencies of the different words in the dataset, it also tells us something more important about the frequent words. Most of them, such as the, a, and and, are not very informative. Just like punctuation, they are not likely to help us distinguish the categories (positive and negative reviews) that we're interested in because they occur very frequently in most documents. These kinds of words are called **stopwords** and are usually removed from natural language data before NLP precisely because they don't add much information.

NLTK provides a list of common stopwords for different languages. Let's take a look at some of the English stopwords:

```
1  import nltk
2  from nltk.corpus import stopwords
3
4  stop_words = list(set(stopwords.words('english')))
5  print(len(stop_words))
6  print(stop_words[0:50])

179
['each', 'did', 'such', "should've", 'can', "shouldn't", 'him', 'as', 'there', 'n
ow', 'out', 'o', 'during', 'because', 'yourselves', 've', 'up', 'ours', 'the', 't
o', 'ma', 'both', 'am', 'after', 'all', 'your', 'hers', "mightn't", 'hadn', 'me',
"that'll", 'isn', 'same', 'from', 'just', 'he', 'she', 'haven', "hasn't", 'on',
'who', 'yourself', 'had', 'whom', 'we', 'down', 'shan', 'i', 'under', 'no']
```

Figure 6.5 – Examples of 50 English stopwords

Figure 6.5 shows the code we can use to examine English stopwords, using the following steps:

1. Import nltk and the stopwords package.

2. Make the set of stopwords into a Python list so that we can do list operations on it.

3. Find out how long the list is with the Python len() function and print the value (179 items).

4. Print the first 50 stopwords.

Stopwords are available for other languages in NLTK as well. We can see the list by running the code in *Figure 6.6*:

```
1  languages = stopwords.fileids()
2  print('Stopwords for ', len(languages), ' languages are included in NLTK')
3  print(languages)

Stopwords for  29  languages are included in NLTK
['arabic', 'azerbaijani', 'basque', 'bengali', 'catalan', 'chinese', 'danish', 'dutch',
'english', 'finnish', 'french', 'german', 'greek', 'hebrew', 'hinglish', 'hungarian', 'in
donesian', 'italian', 'kazakh', 'nepali', 'norwegian', 'portuguese', 'romanian', 'russia
n', 'slovene', 'spanish', 'swedish', 'tajik', 'turkish']
```

Figure 6.6 – NLTK languages with stopwords

The code has the following three steps:

1. Collect the names of the languages with NLTK stopwords. The stopwords are contained in files, one file per language, so getting the files is a way to get the names of the languages.

2. Print the length of the list of languages.

3. Print the names of the languages.

Once we have the list of stopwords, it is easy to remove them from a corpus. After removing the punctuation, as we saw in *Figure 6.3*, we can just iterate through the list of words in the corpus, removing the words that are in the stopwords list, as shown here:

```
words_stop = [w for w in words_no_punct if not w in stop_words]
```

As an aside, it would clearly be more efficient to combine removing punctuation and stopwords with one iteration through the entire list of words, checking each word to see whether it is either punctuation or a stopword, and removing it if it is either one. We show these steps separately for clarity, but you may want to combine these two steps in your own projects. Just as we did in the case of removing punctuation, we can use the NLTK frequency distribution function to see which words other than stopwords are most common, and display them, as shown in *Figure 6.7*:

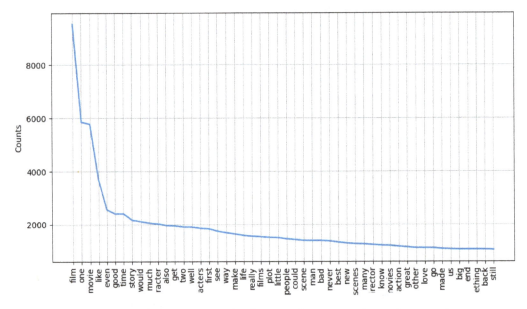

Figure 6.7 – Most common words, after removing punctuation and stopwords

We can see in *Figure 6.7* that removing stopwords gives us a much clearer picture of the words in the movie review corpus. The most common word is now film, and the third most common word is movie. This is what we would expect from a corpus of movie reviews. A similar review corpus (for example, a corpus of product reviews) would be expected to show a different distribution of frequent words. Even looking at the simple information provided by frequency distributions can sometimes be enough to do some NLP tasks.

For example, in authorship studies, we're interested in trying to attribute a document whose author is unknown to a known author. The word frequencies in the document whose author we don't know can be compared to those in the documents with a known author. Another example would be domain classification, where we want to know whether the document is a movie review or a product review.

Visualizing data with Matplotlib, Seaborn, and pandas

While NLTK has some basic plotting functions, there are some other, more powerful Python plotting packages that are used very often for plotting all kinds of data, including NLP data. We'll look at some of these in this section.

Matplotlib (`https://matplotlib.org/`) is very popular. Matplotlib can create a variety of visualizations, including animations and interactive visualizations. We'll use it here to create another visualization of the data that we plotted in *Figure 6.4*. Seaborn (`https://seaborn.pydata.org/`) is built on Matplotlib and is a higher-level interface for producing attractive visualizations. Both of these packages frequently make use of another Python data package, pandas (`https://pandas.pydata.org/`), which is very useful for handling tabular data.

For our example, we'll use the same data that we plotted in *Figure 6.7*, but we'll use Matplotlib, Seaborn, and pandas to create our visualization.

Figure 6.8 shows the code for this example:

```
1  import nltk
2  import pandas as pd
3  import seaborn as sns
4  import matplotlib.pyplot as plt
5
6  frequency_cutoff = 25
7  all_fdist = nltk.FreqDist(freq_without_stopwords).most_common(frequency_cutoff)
8
```

Figure 6.8 – Python code for collecting the 25 most common
words, after removing punctuation and stopwords

Figure 6.8 shows the preparatory code for plotting the data, with the following steps:

1. Import the NLTK, pandas, Seaborn, and Matplotlib libraries.

2. Set the frequency cutoff to 25 (this can be whatever number we're interested in).

3. Compute a frequency distribution of the corpus words without stopwords (line 7).

We start by importing NLTK, pandas, Seaborn, and Matplotlib. We set the frequency cutoff to 25, so that we will plot only the 25 most common words, and we get the frequency distribution of our data at line 7.

The resulting plot is shown in *Figure 6.9*:

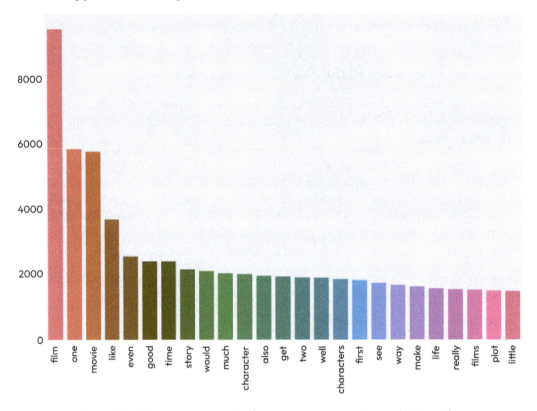

Figure 6.9 – Most common words, after removing punctuation and stopwords,
displayed using Matplotlib, Seaborn, and pandas libraries

Figure 6.7 and *Figure 6.9* show the same data. For example, the most common word in each figure is film, followed by one, and then movie. Why would you pick one visualization tool over the other? One difference is that it seems to be easier to see the information for a specific word in *Figure 6.9*, which is probably because of the bar graph format, where every word is associated with a specific bar. In contrast, in *Figure 6.7*, it is a little harder to see the frequencies of individual words because this isn't very clear from the line.

On the other hand, the overall frequency distribution is easier to see in *Figure 6.7* because of the continuous line. So, choosing one type of visualization over the other is really a matter of the kind of information that you want to emphasize. Of course, there is also no reason to limit yourself to one visualization if you want to see your data in different ways.

As a side note, the pattern in both figures, where we have very high frequencies for the most common words, falling off rapidly for less common words, is very common in natural language data. This pattern illustrates a concept called **Zipf's law** (for more information about this concept, see `https://en.wikipedia.org/wiki/Zipf%27s_law`).

Looking at another frequency visualization technique – word clouds

A **word cloud** is another way of visualizing the relative frequencies of words. A word cloud displays the words in a dataset in different font sizes, with the more frequent words shown in larger fonts. Word clouds are a good way to make frequent words pop out visually.

The code in *Figure 6.10* shows how to import the `WordCloud` package and create a word cloud from the movie review data:

```
1  import nltk
2  from wordcloud import WordCloud
3  import matplotlib.pyplot as plt
4  import pandas as pd
5
6  frequency_cutoff = 200
7  all_fdist = nltk.FreqDist(freq_without_stopwords).most_common(frequency_cutoff)
8  all_fdist = pd.Series(dict(all_fdist))
9
10 long_words = dict([(m, n) for m, n in all_fdist.items() if len(m) > 2])
11 wordcloud = WordCloud(colormap="tab10",background_color="white").generate_from_frequencies(long_words)
12 plt.figure( figsize=(20,15))
13 plt.imshow(wordcloud, interpolation='bilinear')
14 plt.axis("off")
15 plt.show()
```

Figure 6.10 – Python code for most common words, shown as a word cloud

The code shows the following steps:

1. We import a new library, `WordCloud`, at line 2. For this display, we will select only the most *common 200 words* (with a frequency cutoff of `200` at line 6).

2. We create a frequency distribution at line 7.

3. The frequency distribution is converted to a pandas Series at line 8.

4. To reduce the number of very short words, we include only words that are longer than two letters at line 10.

5. The code in line 11 generates the word cloud. The `colormap` parameter specifies one of the many Matplotlib color maps (color maps are documented at `https://matplotlib.org/stable/tutorials/colors/colormaps.html`). You can experiment with different color schemes to find the ones that you prefer.

6. Lines 12-14 format the plot area.

7. Finally, the plot is displayed at line 15.

Figure 6.11 shows the word cloud that results from the code in *Figure 6.10*:

Figure 6.11 – Word cloud for word frequencies in movie reviews

As we saw in *Figure 6.7* and *Figure 6.9*, the most common words are film, one, and movie. However, the word cloud visualization makes the most common words stand out in a way that the graphs do not. On the other hand, the less frequent words are difficult to differentiate in the word cloud. For example, it's hard to tell whether good is more frequent than story from the word cloud. This is another example showing that you should consider what information you want to get from a visualization before selecting one.

The next section will show you how to dig into the data a little more deeply to get some insight into subsets of the data such as positive and negative reviews.

Positive versus negative movie reviews

We might want to look at differences between the different classes (positive and negative) of movie reviews, or, more generally, any different categories. For example, are the frequent words in positive and negative reviews different? This preliminary look at the properties of the positive and negative reviews can give us some insight into how the classes differ, which, in turn, might help us select the approaches that will enable a trained system to differentiate the categories automatically.

We can use any of the visualizations based on frequency distributions for this, including the line graph in *Figure 6.7*, the bar graph in *Figure 6.9*, or the word cloud visualization in *Figure 6.11*. We'll use the word cloud in this example because it is a good way of seeing the differences in word frequencies between the two frequency distributions.

Looking at the code in *Figure 6.12*, we start by importing the libraries we'll need:

```
1  import nltk
2  from nltk.corpus import movie_reviews
3  from nltk.corpus import stopwords
4  from wordcloud import WordCloud
5  import matplotlib.pyplot as plt
6
```

Figure 6.12 – Importing libraries for computing word clouds

The next thing we'll do is create two functions that will make it easier to perform similar operations on different parts of the corpus:

```
7  # remove punctuation and stopwords
8  def clean_corpus(corpus):
9      cleaned_corpus = []
10     for word in corpus:
11         if word.isalnum() and not word in stop_words:
12             cleaned_corpus.append(word)
13     return(cleaned_corpus)
14
15 # show a word cloud, given a frequency distribution
16 def plot_freq_dist(freq_dist):
17     frequency_cutoff = 50
18     long_words = dict([(m, n) for m, n in freq_dist.items() if len(m) > 2])
19     wordcloud = WordCloud(colormap="tab10",background_color="white").generate_from_frequencies(long_words)
20     plt.figure( figsize=(20,15))
21     plt.imshow(wordcloud, interpolation='bilinear')
22     plt.axis("off")
23     plt.show()
```

Figure 6.13 – Functions for computing word clouds

The first function in *Figure 6.13*, clean_corpus(), removes the punctuation and stopwords from a corpus. The second one, plot_freq_dist(), plots the word cloud from the frequency distribution. Now we're ready to create the word clouds:

```
25 stop_words = list(set(stopwords.words('english')))
26 corpus_neg_words = movie_reviews.words(categories="neg")
27 corpus_pos_words = movie_reviews.words(categories="pos")
28 negative_words = clean_corpus(corpus_neg_words)
29 positive_words = clean_corpus(corpus_pos_words)
30 neg_freq = nltk.FreqDist(negative_words)
31 pos_freq = nltk.FreqDist(positive_words)
32
33 plot_freq_dist(pos_freq)
```

Figure 6.14 – The code for displaying word frequencies for positive and negative reviews in a word cloud

We create the word cloud with the following steps:

1. Now that we've defined the functions, we separate the corpus into positive and negative reviews. This is done by the code shown in lines 28 and 29.

2. Instead of asking for all the words in the corpus, as we saw in *Figure 6.2*, we ask for words in specific categories. In this example, the categories are positive (pos) and negative (neg).

3. We remove the stopwords and punctuation in each set of words in lines 28 and 29.

4. We then create frequency distributions for the positive and negative words in lines 30 and 31. Finally, we plot the word cloud.

Figure 6.15 shows the word cloud for the word frequencies in the positive reviews:

Figure 6.15 – Displaying word frequencies for positive reviews in a word cloud

Figure 6.16 shows the word frequencies for negative reviews in a word cloud:

Figure 6.16 – Displaying word frequencies for negative reviews in a word cloud

Comparing *Figures 6.15* and *6.16* with the original word cloud in *Figure 6.11*, we can see that the words `film`, `one`, and `movie` are the most frequent words in positive and negative reviews, as well as being the most frequent overall words, so they would not be very useful in distinguishing positive and negative reviews.

The word `good` is larger (and therefore more frequent) in the positive review word cloud than in the negative review word cloud, which is just what we would expect. However, `good` does occur fairly frequently in the negative reviews, so it's not a definite indication of a positive review by any means. Other differences are less expected – for example, `story` is more common in the positive reviews, although it does occur in the negative reviews. This comparison shows that the problem of distinguishing between positive and negative reviews is unlikely to be solved by simple keyword-spotting techniques.

The techniques we'll be looking at in *Chapter 9* and *Chapter 10*, will be more suitable for addressing the problem of classifying texts into different categories. However, we can see that this initial exploration with word clouds was very useful in eliminating a simple keyword-based approach from the set of ideas we are considering.

In the next section, we'll look at other frequencies in the corpus.

Looking at other frequency measures

So far we've only looked at word frequencies, but we can look at the frequency of any other text property we can measure. For example, we can look at the frequencies of different characters or different parts of speech. You can try extending the code that was presented earlier in this chapter, in the previous section, *Frequency distributions*, to count some of these other properties of the movie review texts.

One important property of language is that words don't just occur in isolation – they occur in specific combinations and orders. The meanings of words can change dramatically depending on their contexts. For example, *not a good movie* has a very different meaning (in fact, the opposite meaning) from *a good movie*. There are a number of techniques for taking word context into account, which we will explore in *Chapters 8 to 11*.

However, here we'll just describe one very simple technique – looking at words that occur next to each other. Where two words occur together, this is called a **bigram**. An example is *good movie*. A set of three words in sequence is called a **trigram**, and in general, any number of words in a sequence is called an **ngram**. NLTK provides a function for counting ngrams, `ngrams()`, which takes the desired value of n as an argument. *Figure 6.17* shows code for counting and displaying ngrams in the movie review corpus:

```
1  import nltk
2  from nltk.util import ngrams
3  from nltk.corpus import movie_reviews
4
5  frequency_cutoff = 25
6
7  # collect the words from the corpus
8  corpus_words = movie_reviews.words()
9
10 # remove punctuation and stopwords
11 cleaned_corpus = clean_corpus(corpus_words)
```

Figure 6.17 – Computing bigrams in the movie review data

The code shows the following initialization steps:

1. We start by importing `nltk`, the `ngrams` function, and the movie review corpus.

2. We set the frequency cutoff to 25, but as in previous examples, this can be any number that we think will be interesting.

3. We collect all the words in the corpus at line 8 (if we wanted just the words from the positive reviews or just the words from the negative reviews, we could get those by setting `categories = 'neg'` or `categories = 'pos'`, as we saw in *Figure 6.14*).

4. Finally, we remove punctuation and stopwords at line 11, using the `clean_corpus` function defined in *Figure 6.13*.

In *Figure 6.18*, we collect the bigrams:

```
13  # collect the bigrams in the corpus
14  bigrams = ngrams(cleaned_corpus,2)
15
16  # make a list from the bigrams
17  list_bigrams = list(bigrams)
18
19  # put together the bigrams into a single string
20  consolidated_bigrams = []
21  for bigram in list_bigrams:
22      consolidated_bigram = bigram[0] + " " + bigram[1]
23      consolidated_bigrams.append(consolidated_bigram)
```

Figure 6.18 – Computing bigrams in the movie review data

The bigrams are collected using the following steps:

1. The `ngrams()` function is used at line 14, with an argument, 2, indicating that we want bigrams (or pairs of two adjacent words). Any number can be used here, but very large numbers are not likely to be very useful. This is because as the value of n increases, there will be fewer and fewer ngrams with that value. At some point, there won't be enough examples of a particular ngram to provide any information about patterns in the data. Bigrams or trigrams are usually common enough in the corpus to be helpful in identifying patterns.

2. In lines 21-23, we loop through the list of bigrams, creating a string from each pair.

    In *Figure 6.19*, we make a frequency distribution of the bigrams and display it.

```
25  # make a frequency distribution from the bigrams
26
27  freq_bigrams = nltk.FreqDist(consolidated_bigrams).most_common(frequency_cutoff)
28  # Convert to a Pandas series
29  all_fdist = pd.Series(dict(freq_bigrams))
30
31  # set figure and axis variables and set sizes for the x and y axes
32  fig, ax = plt.subplots(figsize=(50,40))
33
34  # create a bar graph using Seaborn
35  sns.set(font_scale=2)
36  # display the bigrams on the y-axis and the counts on the x-axis
37  all_plot = sns.barplot(x=all_fdist.values, y=all_fdist.index, ax=ax)
38
39  plt.show()
```

Figure 6.19 – Displaying bigrams in the movie review data

3. We use the familiar NLTK `FreqDist()` function on our list of bigrams and convert it to a pandas Series in line 29.

4. The rest of the code sets up a bar graph plot in Seaborn and Matplotlib, and finally, displays it at line 39.

Since we're using a frequency cutoff of 25, we'll only be looking at the most common 25 bigrams. You may want to experiment with larger and smaller frequency cutoffs.

Figure 6.20 shows the results of the plot that we computed in *Figure 6.19*. Because bigrams, in general, will be longer than single words and take more room to display, the *x* and *y* axes have been swapped around so that the counts of bigrams are displayed on the *x* axis and the individual bigrams are displayed on the *y* axis:

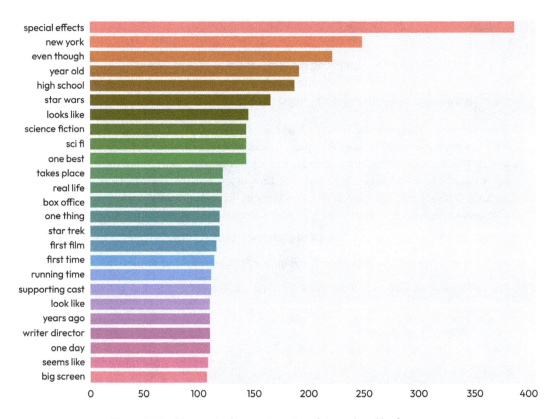

Figure 6.20 – Bigrams in the movie review data, ordered by frequency

Figure 6.20 reveals several interesting facts about the corpus. For example, the most common bigram, `special effects`, occurs about 400 times in the corpus, compared to the most common single word, `film`, which occurs nearly 8,000 times. This is to be expected because two words have to occur together to count as a bigram. We also see that many of the bigrams are **idioms**. An idiom is a combination of two or more words together that have a meaning that is not simply the combination of the meanings of the individual words. `New York` and `Star Trek` are examples in this list. Other bigrams are not idioms but just common phrases, such as `real life` and `one day`. In this list, all of the bigrams are very reasonable, and it is not surprising to see any of them in a corpus of movie reviews.

As an exercise, try comparing the bigrams in the positive and negative movie reviews. In looking at single-word frequencies, we saw that the most common words were the same in both positive and negative reviews. Is that also the case with the most common bigrams?

This section covered some ways to get insight into our datasets by simple measurements such as counting words and bigrams. There are also some useful exploratory techniques for measuring and visualizing the similarities among documents in a dataset. We will cover these in the next section.

Measuring the similarities among documents

So far, we've been looking at visualizing the frequencies of various properties of the corpus, such as words and bigrams, with tools such as line charts, bar graphs, and word clouds. It is also very informative to visualize document similarity – that is, how similar documents in a dataset are to each other. There are many ways to measure document similarity, which we will discuss in detail in later chapters, especially *Chapter 9*, *Chapter 10*, *Chapter 11*, and *Chapter 12*. We will start here with an introductory look at two basic techniques.

BoW and k-means clustering

For now, we will use a very simple idea, called **bag of words** (**BoW**). The idea behind BoW is that documents that are more similar to each other will contain more of the same words. For each document in a corpus and for each word in the corpus, we look at whether or not that word occurs in that document. The more words that any two documents have in common, the more similar to each other they are. This is a very simple measurement, but it will give us a basic document similarity metric that we can use to illustrate visualizations of document similarity.

The code in *Figure 6.21* computes the BoW for the movie review corpus. You do not need to be concerned with the details of this code at this time, since it's just a way of getting a similarity measure for the corpus. We will use this similarity metric (that is, BoW) to see how similar any two documents are to each other. The BoW metric has the advantage that it is easy to understand and compute. Although it is not the state-of-the-art way of measuring document similarity, it can be useful as a quick first step:

```
1  import random
2
3  #import the training data
4  from nltk.corpus import movie_reviews
5  corpus_words = movie_reviews.words()
6
7  # remove punctuation and stopwords
8  cleaned_corpus = clean_corpus(corpus_words)
9
10 all_words = nltk.FreqDist(w for w in cleaned_corpus)
11 max_words = 1000
12 word_features = list(all_words)[:max_words]
```

Figure 6.21 – Setting up the BoW computation

Figure 6.21 shows the process of getting the most frequent 1,000 words from the corpus and making them into a list. The number of words we want to keep in the list is somewhat arbitrary – a very long list will slow down later processing and will also start to include rare words that don't provide much information.

The next steps, shown in *Figure 6.22*, are to define a function to collect the words in a document and then make a list of the documents:

```
13
14 def document_features(document):
15     features = {}
16     for word in word_features:
17         if word in document:
18             features[word] = 1
19         else:
20             features[word] = 0
21     return features
22
23 # make a list of documents
24 documents = [(list(movie_reviews.words(fileid)), category)
25               for category in movie_reviews.categories()
26               for fileid in movie_reviews.fileids(category)]
27
28 random.shuffle(documents)
```

Figure 6.22 – Collecting the words that occur in a document

The `document_features()` function in *Figure 6.22* iterates through the given document creating a Python dictionary with the words as keys and 1s and 0s as the values, depending on whether or not the word occurred in the document. Then, we create a list of the features for each document and display the result in *Figure 6.23*:

```
30  # collect features, that is, words that occur in a document
31  featuresets = [(document_features(document), category) for (document,category) in documents]
32
33  #remove categories for display
34  docnumber = 0
35  new_featuresets = {}
36  for featureset in featuresets:
37      new_featureset = featureset[0]
38      new_featuresets[docnumber]= new_featureset
39      docnumber += 1
40
41  # display the words that occur in the first 10 documents, the bag of words
42  df_featuresets = pd.DataFrame.from_dict(data = new_featuresets, orient = 'index', columns = word_features)
43  df_featuresets.head(10)
```

Figure 6.23 – Computing the full feature set for all documents and displaying the resulting BoW

Although the list of features for each document includes its category, we don't need the category in order to display the BoW, so we remove it at lines 34-39.

We can see the first 10 documents in the resulting BoW itself in *Figure 6.24*:

	film	one	movie	like	even	good	time	story	would	much	...	spielberg	development	etc	language	blue	proves	vampire	seemingly	basic	caught
0	1	0	1	1	0	0	0	1	1	1	...	0	0	0	0	0	0	0	0	0	0
1	1	1	1	1	1	0	1	1	1	1	...	1	0	0	0	0	0	0	0	0	1
2	1	1	1	1	0	1	0	0	0	1	...	0	0	0	0	0	0	0	0	0	0
3	0	1	1	1	1	0	0	1	0	1	...	0	0	0	0	0	0	0	0	0	0
4	1	0	1	0	0	1	0	1	0	0	...	0	0	0	0	0	0	0	0	0	0
5	1	1	1	1	1	1	1	0	0	0	...	0	0	0	0	0	0	0	0	1	0
6	1	1	1	0	1	1	1	0	1	1	...	0	0	0	0	0	0	0	0	0	1
7	1	1	1	0	1	1	1	1	0	0	...	0	1	0	0	0	0	0	0	0	0
8	1	1	1	1	1	1	1	1	1	1	...	0	0	0	0	0	0	0	0	0	0
9	1	1	1	0	1	1	1	1	0	0	...	0	0	0	0	0	0	0	0	0	1

10 rows × 1000 columns

Figure 6.24 – BoW for the movie review corpus

In *Figure 6.24*, each of the 10 rows of 0s and 1s represents one document. There is one column for each word in the corpus. There are 1,000 columns, but this is too many to display, so we only see the first few and last few columns. The words are sorted in order of frequency, and we can see that the most frequent words (`film`, `one`, and `movie`) are the same words that we found to be the most frequent (except for stopwords) in our earlier explorations of word frequencies.

Each document is represented by a row in the BoW. This list of numbers in these rows is a **vector**, which is a very important concept in NLP, which we will be seeing much more of in later chapters. Vectors are used to represent text documents or other text-based information as numbers. Representing words as numbers opens up many opportunities for analyzing and comparing documents, which are difficult to do when the documents are in text form. Clearly, BoW loses a lot of information compared to the original text representation – we no longer can tell which words are even close to each other in the text, for example – but in many cases, the simplicity of using BoW outweighs the disadvantage of losing some of the information in the text.

One very interesting way that we can use vectors in general, including BoW vectors, is to try to capture document similarities, which can be a very helpful first step in exploring a dataset. If we can tell which documents are similar to which other documents, we can put them into categories based on their similarity. However, just looking at the document vectors in the rows of *Figure 6.24* does not provide much insight, because it's not easy to see any patterns. We need some tools for visualizing document similarity.

A good technique for visualizing document similarities is k-means clustering. **K-means clustering** tries to fit documents into visual clusters based on their similarities. In our case, the similarity metric that we will be using is BoW, where the assumption is that the more words two documents have in common, the more similar they are. K-means clustering is an iterative algorithm that represents similarities visually by the distance between items – more similar items are closer together in space. The k value refers to the number of clusters we want to have and is selected by the developer. In our case, we will start with 2 as the value of k since we have 2 known classes, corresponding to the sets of positive and negative reviews.

Figure 6.25 shows the code for computing and displaying the results of k-means clustering on the BoW we computed in *Figure 6.23*. We do not have to go into detail about the code in *Figure 6.25* because it will be covered in detail in *Chapter 12*. However, we can note that this code uses another important Python machine learning library, `sklearn`, which is used to compute the clusters:

```
1  import numpy as np
2  import matplotlib.pyplot as plt
3
4  from sklearn.cluster import KMeans
5  from sklearn.decomposition import TruncatedSVD
6
7  true_k = 2
```

Figure 6.25 – Setting up for k-means clustering

The first step is to import the libraries we'll need and set the number of clusters we want (`true_k`).

The next part of the code, shown in *Figure 6.26*, computes the k-means clusters.

```
9  # truncatedSVD for reducing dimensions to 2 for display
10 truncatedSVD = TruncatedSVD(n_components = 2)
11 X_2D = truncatedSVD.fit_transform(df_featuresets)
12
13 kmeans = KMeans(n_clusters = true_k,
14                 init='k-means++',
15                 max_iter=100, # Maximum iterations
16                 n_init=10)  # Number of times to run the k-means algorithm
17
18 result = kmeans.fit(X_2D)
19 labels = result.labels_
```

Figure 6.26 – K-means clustering for visualizing document similarities based on the BoW

The steps for computing the clusters as shown in *Figure 6.26* are as follows:

1. Reduce the dimensions to 2 for display.

2. Initialize a kmeans object (line 13).

3. Compute the result (line 18).

4. Get the labels from the result.

The final step is plotting the clusters:

```
21 cm = plt.get_cmap('Accent')
22
23 # plot clusters in different colors
24 for cluster in range(true_k):
25     current_color = cm(1.*cluster/(true_k))
26     plt.scatter(X_2D[labels == cluster, 0], X_2D[labels == cluster, 1],
27             color = current_color, label='cluster ' + str(cluster))
28
29 plt.rcParams["figure.figsize"] = (20,20)
30 plt.rcParams['font.size'] = '12'
31 plt.show()
```

Figure 6.27 – Plotting k-means clustering

Figure 6.27 plots the clusters by iterating through the clusters and printing the items in each cluster in the same color.

The results of using k-means clustering on the movie review BoW can be seen in *Figure 6.28*:

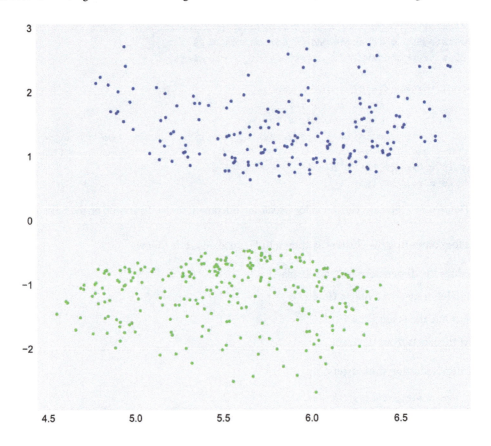

Figure 6.28 – K-means clustering for the movie corpus with two classes

There are two major clusters in *Figure 6.28*, one above and one below the 0 point on the *y* axis. The colors are defined based on the Matplotlib Accent color map at line 21. (More information about the many color maps available in Matplotlib can be found at https://matplotlib.org/stable/tutorials/colors/colormaps.html.)

Each dot in *Figure 6.28* represents one document. Because the clusters are clearly separated, we can have some confidence that the similarity metric (BoW) reflects some real difference between the two classes of documents. However, that does not mean that the most insightful number of classes for this data is necessarily two. It is always worth checking out other numbers of classes; that is, other values of k. This can easily be done by changing the value of true_k at line 7 in *Figure 6.25*. For example, if we change the value of true_k to 3, and thereby specify that the data should be divided into three classes, we'll get a chart like the one in *Figure 6.29*:

Figure 6.29 – K-means clustering for the movie corpus with three classes

There are definitely three clear classes in *Figure 6.29*, although they aren't as nicely separated as the classes in *Figure 6.28*. This could mean that the two classes of positive and negative reviews don't tell the whole story. Perhaps there actually should be a third class of *neutral* reviews? We could investigate this by looking at the documents in the three clusters, although we won't go through that exercise here.

By comparing the results in *Figures 6.28* and *6.29*, we can see that initial data exploration can be very useful in deciding how many classes in which to divide the dataset at the outset. We will take up this topic in much greater detail in *Chapter 12*.

We have so far seen a number of ways of visualizing the information in a dataset, including word and bigram frequencies, as well as some introductory visualizations of document similarity.

Let's now consider some points about the overall process of visualization.

General considerations for developing visualizations

Stepping back a bit from the specific techniques we have reviewed so far, we will next discuss some general considerations about visualizations. Specifically, in the next sections, we will talk about what to measure, followed by how to represent these measurements and the relationships among the measurements. Because the most common visualizations are based on representing information in the *XY* plane in two dimensions, we will mainly focus on visualizations in this format, starting with selecting among measurements.

Selecting among measurements

Nearly all NLP begins with measuring some property or properties of texts we are analyzing. The goal of this section is to help you understand the different kinds of text measurements that are available in NLP projects.

So far, we've primarily focused on measurements involving words. Words are a natural property to measure because they are easy to count accurately – in other words, counting words is a **robust** measurement. In addition, words intuitively represent a natural aspect of language. However, just looking at words can lead to missing important properties of the meanings of texts, such as those that depend on considering the order of words and their relationship to other words in the text.

To try to capture richer information that does take into account the orders of words and their relationships, we can measure other properties of texts. We can, for example, count characters, syntactic constructions, parts of speech, ngrams (sequences of words), mentions of named entities, and word lemmas.

As an example, we could look at whether pronouns are more common in positive movie reviews than in negative movie reviews. However, the downside of looking at these richer properties is that, unlike counting words, measurements of richer properties are less robust. That means they are more likely to include errors that would make the measurements less accurate. For example, if we're counting verbs by using the results of part of speech tagging, an incorrect verb tag that should be a noun tag would make the verb count incorrect.

For these reasons, deciding what to measure is not always going to be a cut-and-dry decision. Some measurements, such as those that are based on words or characters, are very robust but exclude information that could be important. Other measurements, such as counting parts of speech, are less robust but more informative. Consequently, we can't provide hard and fast rules for deciding what to measure. However, one rule of thumb is to start with the simplest and most robust approaches, such as counting words, to see whether your application is working well with those techniques. If it is, you can stick with the simplest approaches. If not, you can try using richer information. You should also keep in mind that you aren't limited to only one measurement.

Once you have selected what you want to measure, there are other general considerations that have to do with visualizing your measurements. In the next sections, we will talk about representing variables on the axes of the *XY* plane, different kinds of scales, and dimensionality reduction.

Insights from representing independent and dependent variables

Most measurement involves measuring one **independent variable** that we have control over and measuring another variable that we do not control (a **dependent variable**) and seeing what the relationship is. For example, in *Figure 6.4*, the words in the dataset are the independent variable, on the *x* axis, and their counts are the dependent variable, on the *y* axis. Going further, suppose we want to evaluate a hypothesis that we can tell that a review is positive if it contains the word good and we can tell that it's negative if it contains bad. We can test this hypothesis by looking at the counts for good and bad (the dependent variable) for positive and negative reviews (the independent variable), as shown in *Figure 6.30*:

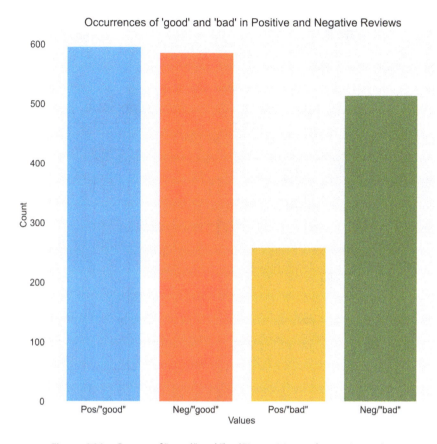

Figure 6.30 – Counts of "good" and "bad" in positive and negative reviews

The columns on the *x* axis represent positive reviews containing the word good, negative reviews containing the word good, positive reviews containing the word bad, and negative reviews containing the word bad, respectively. Clearly, the hypothesis that good signals a positive review and bad signals a negative review is wrong. In fact, good occurs more often than bad in negative reviews. This kind of exploration and visualization can help us rule out lines of investigation at the outset of our project that are unlikely to be fruitful. In general, bar charts such as the one in *Figure 6.30* are a good way to represent categorical independent variables or data that occurs in distinct classes.

After looking at displaying the values of dependent and independent variables in graphs like *Figure 6.30*, we can turn to another general consideration: linear versus log scales.

Log scales and linear scales

Sometimes you will see a pattern, as in *Figures 6.4* and *6.7*, where the first few items on the *x* axis have extremely high values and where the values of the other items drop off quickly. This makes it hard to see what's going on in the part of the graph on the right where the items have lower values. If we see this pattern, it suggests that a **log scale** may be a better choice for the *y* axis than the **linear scale** in *Figure 6.2*. We can see a log scale for the data from *Figure 6.7* in *Figure 6.31*:

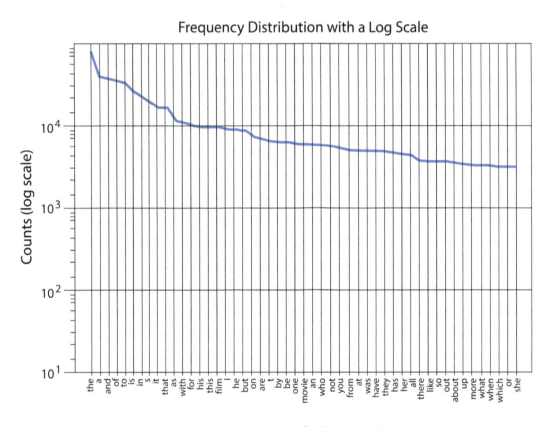

Figure 6.31 – Log scale visualization for the data in Figure 6.4

Figure 6.31 shows a log scale plot of the data in *Figure 6.7*. The *y*-axis values are the counts of the 50 most common words in equally spaced powers of 10, so each value on the *y* axis is 10 times greater than the previous one. Compared with *Figure 6.7*, we can see that the line representing the counts of frequencies is much flatter, especially toward the right. In the log display, it is easier to compare the frequencies of these less common, but still very frequent, words. Because even the least frequent word in this chart, she, still occurs over 3,000 times in the dataset, the line does not go below 10^3.

You should keep the option of displaying your data in a log scale in mind if you see a pattern like the one in *Figure 6.7*.

So far, we've been looking at data in two dimensions that can easily be displayed on paper or a flat screen. What happens if we have more than two dimensions? Let's now consider higher dimensional data.

Dimensions and dimensionality reduction

The examples we've seen up to this point have been largely two-dimensional diagrams with x and y axes. However, many NLP techniques result in data that is of much higher dimensionality. While two dimensions are easy to plot and visualize, higher dimensional data can be harder to understand. Three-dimensional data can be displayed by adding a z axis and rotating the graph so that the x and z axes are at 45-degree angles on the display. A fourth or time dimension can be added by animating an on-screen graph and showing the changes over time. However, dimensions beyond four are really impossible for humans to visualize, so there is no way to display them in a meaningful way.

However, in many cases, it turns out that some of the higher dimensions in NLP results can be removed without seriously impacting either the visualization or the information produced by NLP. This is called **dimensionality reduction**. Dimensionality reduction techniques are used to remove the less important dimensions from the data so that it can be displayed more meaningfully. We will look at dimensionality reduction in more detail in *Chapter 12*.

Note that dimensionality reduction was done in *Figures 6.28* and *6.29* in order to display the results on a two-dimensional page.

The final topic in this chapter addresses some things we can learn from visualizations and includes some suggestions for using visualization to decide how to proceed with the next phases of an NLP project.

Using information from visualization to make decisions about processing

This section includes guidance about how visualization can help us make decisions about processing. For example, in making a decision about whether to remove punctuation and stopwords, exploring word frequency visualizations such as frequency distribution and word clouds can tell us whether very common words are obscuring patterns in the data.

Looking at frequency distributions of words for different categories of data can help rule out simple keyword-based classification techniques.

Frequencies of different kinds of items, such as words and bigrams, can yield different insights. It can also be worth exploring the frequencies of other kinds of items, such as parts of speech or syntactic categories such as noun phrases.

Displaying document similarities with clustering can provide insight into the most meaningful number of classes that you would want to use in dividing a dataset.

The final section summarizes the information that we learned in this chapter.

Summary

In this chapter, we learned about some techniques for the initial exploration of text datasets. We started out by exploring data by looking at the frequency distributions of words and bigrams. We then discussed different visualization approaches, including word clouds, bar graphs, line graphs, and clusters. In addition to visualizations based on words, we also learned about clustering techniques for visualizing similarities among documents. Finally, we concluded with some general considerations for developing visualizations and summarized what can be learned from visualizing text data in various ways. The next chapter will cover how to select approaches for analyzing NLU data and two kinds of representations for text data – symbolic representations and numerical representations.

7
Selecting Approaches and Representing Data

This chapter will cover the next steps in getting ready to implement a **natural language processing** (**NLP**) application. We start with some basic considerations about understanding how much data is needed for an application, what to do about specialized vocabulary and syntax, and take into account the need for different types of computational resources. We then discuss the first steps in NLP – text representation formats that will get our data ready for processing with NLP algorithms. These formats include symbolic and numerical approaches for representing words and documents. To some extent, data formats and algorithms can be mixed and matched in an application, so it is helpful to consider data representation independently from the consideration of algorithms.

The first section will review general considerations for selecting NLP approaches that have to do with the type of application we're working on, and with the data that we'll be using.

In this chapter, we will cover the following topics:

- Selecting NLP approaches
- Representing language for NLP applications
- Representing language numerically with vectors
- Representing words with context-independent vectors
- Representing words with context-dependent vectors

Selecting NLP approaches

NLP can be done with a wide variety of possible techniques. When you get started on an NLP application, you have many choices to make, which are affected by a large number of factors. One of the most important factors is the type of application itself and the information that the system needs to extract from the data to perform the intended task. The next section addresses how the application affects the choice of techniques.

Fitting the approach to the task

Recall from *Chapter 1*, that there are many different types of NLP applications divided into **interactive** and **non-interactive applications**. The type of application you choose will play an important role in choosing the technologies that will be applied to the task. Another way of categorizing applications is in terms of the level of detail required to extract the needed information from the document. At the coarsest level of analysis (for example, classifying documents into two different categories), techniques can be less sophisticated, faster to train, and less computationally intensive. On the other hand, if the task is training a chatbot or voice assistant that needs to pull out multiple entities and values from each utterance, the analysis needs to be more sensitive and fine-grained. We will see some specific examples of this in later sections of this chapter.

In the next section, we will discuss how data affects our choice of techniques.

Starting with the data

NLP applications are built on datasets or sets of examples of the kinds of data that the target system will need to process. To build a successful application, having the right amount of data is imperative. However, we can't just specify a single number of examples for every application, because the right amount of data is going to be different for different kinds of applications. Not only do we have to have the right amount of data, but we have to have the right kind of data. We'll talk about these considerations in the next two sections.

How much data is enough?

In *Chapter 5*, we discussed many methods of obtaining data, and after going through *Chapter 5*, you should have a good idea of where your data will be coming from. However, in that chapter, we did not address the question of how to tell how much data is needed for your application to accomplish its goal. If there are hundreds or thousands of different possible classifications of documents in a task, then we will need a sufficient number of examples of each category for the system to be able to tell them apart. Obviously, the system can't detect a category if it hasn't ever seen an example of that category, but it will also be quite difficult to detect categories where it has seen very few examples.

If there are many more examples of some classes than others, then we have an **imbalanced** dataset. Techniques for balancing classes will be discussed in detail in *Chapter 14*, but basically, they include **undersampling** (where some items in the more common class are discarded), **oversampling** (where items in the rarer classes are duplicated), and **generation** (where artificial examples from the rarer classes are generated through rules).

Systems generally perform better if they have more data, but the data also has to be representative of the data that the system will encounter at the time of testing, or when the system is deployed as an application. If a lot of new vocabulary has been added to the task (for example, if a company's chatbot has to deal with new product names), the training data needs to be periodically updated for the best performance. This is related to the general question of specialized vocabulary and syntax since product names are a kind of specialized vocabulary. In the next section, we will discuss this topic.

Specialized vocabulary and syntax

Another consideration is how similar the data is to the rest of the natural language we will be processing. This is important because most NLP processing makes use of models that were derived from previous examples of the language. The more similar the data to be analyzed is to the rest of the language, the easier it will be to build a successful application. If the data is full of specialized jargon, vocabulary, or syntax, then it will be hard for the system to generalize from its original training data to the new data. If the application is full of specialized vocabulary and syntax, the amount of training data will need to be increased to include this new vocabulary and syntax.

Considering computational efficiency

The computer resources that are needed to implement a particular NLP approach are an important consideration in selecting an approach. Some approaches that yield good results when tested on laboratory benchmarks can be impractical in applications that are intended for deployment. In the next sections, we will discuss the important consideration of the time required to execute the approaches, both at training time and inference time.

Training time

Some modern neural net models are very computationally intensive and require very long training periods. Even before the actual neural net training starts, there may need to be some exploratory efforts aimed at identifying the best values for hyperparameters. **Hyperparameters** are training parameters that can't directly be estimated during the training process, and that have to be set by the developer. When we return to machine learning techniques in *Chapters 9*, *10*, *11*, and *12*, we will look at specific hyperparameters and talk about how to identify good values.

Inference time

Another important consideration is **inference time**, or the processing time required for a trained system to perform its task. For interactive applications such as chatbots, inference time is not normally a concern because today's systems are fast enough to keep up with a user in an interactive application. If a system takes a second or two to process a user's input, that's acceptable. On the other hand, if the system needs to process a large amount of existing online text or audio data, the inference time should be as fast as possible. For example, Statistica.com (`https://www.statista.com/statistics/259477/hours-of-video-uploaded-to-youtube-every-minute/`) estimated in February 2020 that 500 hours of videos were uploaded to YouTube every minute. If an application was designed to process YouTube videos and needed to keep up with that volume of audio, it would need to be very fast.

Initial studies

Practical NLP requires fitting the tool to the problem. When there are new advances in NLP technology, there can be very enthusiastic articles in the press about what the advances mean. But if you're trying to solve a practical problem, trying to use new techniques can be counterproductive because the newest technologies might not scale. For example, new techniques might provide higher accuracy, but at the cost of very long training periods or very large amounts of data. For this reason, it is recommended that when you're trying to solve a practical problem, do some initial exploratory studies with simpler techniques to see whether they'll solve the problem. Only if the simpler techniques don't address the problem's requirements should more advanced techniques be utilized.

In the next section, we will talk about one of the important choices that need to be made when you design an NLP application – how to represent the data. We'll look at both symbolic and numerical representations.

Representing language for NLP applications

For computers to work with natural language, it has to be represented in a form that they can process. These representations can be **symbolic**, where the words in a text are processed directly, or **numeric**, where the representation is in the form of numbers. We will describe both of these approaches here. Although the numeric approach is the primary approach currently used in NLP research and applications, it is worth becoming somewhat familiar with the ideas behind symbolic processing.

Symbolic representations

Traditionally, NLP has been based on processing the words in texts directly, as words. This approach was embodied in a standard approach where the text was analyzed in a series of steps that were aimed at converting an input consisting of unanalyzed words into a meaning. In a traditional NLP pipeline, shown in *Figure 7.1*, each step in processing, from input text to meaning, produces an output that adds more structure to its input and prepares it for the next step in processing. All of these results are symbolic – that is, non-numerical. In some cases, the results might include probabilities, but the actual results are symbolic:

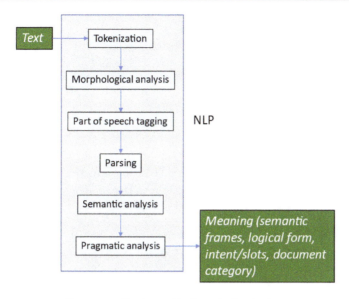

Figure 7.1 – Traditional NLP symbolic pipeline

Although we won't review all of the components of the symbolic approach to processing, we can see a couple of these symbolic results in the following code samples, showing part of speech tagging results and parsing results, respectively. We will not address semantic analysis or pragmatic analysis here since these techniques are generally applied only for specialized problems:

```
import nltk
from nltk.tokenize import word_tokenize
from nltk.corpus import movie_reviews
example_sentences = movie_reviews.sents()
example_sentence = example_sentences[0]
nltk.pos_tag(example_sentence)
```

The preceding code snippet shows the process of importing the movie review database, which we have seen previously, followed by the code for selecting the first sentence and part of speech tagging it. The next code snippet shows the results of part of speech tagging:

```
[('plot', 'NN'),
 (':', ':'),
 ('two', 'CD'),
 ('teen', 'NN'),
 ('couples', 'NNS'),
 ('go', 'VBP'),
 ('to', 'TO'),
 ('a', 'DT'),
```

```
  ('church', 'NN'),
  ('party', 'NN'),
  (',', ','),
  ('drink', 'NN'),
  ('and', 'CC'),
  ('then', 'RB'),
  ('drive', 'NN'),
  ('.', '.')]
```

The tags (NN, CD, NNS, etc.) shown in the preceding results are those used by NLTK and are commonly used in NLP. They are originally based on the Penn Treebank tags (*Building a Large Annotated Corpus of English: The Penn Treebank* (Marcus et al., CL 1993)).

Another important type of symbolic processing is **parsing**. We can see the result of parsing the first sentence, plot: two teen couples go to a church party, drink and then drive, which we saw in the preceding code snippet, in the following code:

```
import spacy
text = "plot: two teen couples go to a church party, drink and then
drive."
nlp = spacy.load("en_core_web_sm")
doc = nlp(text)
for token in doc:
    print (token.text, token.tag_, token.head.text, token.dep_)

plot NN plot ROOT
: : plot punct
two CD couples nummod
teen NN couples compound
couples NNS go nsubj
go VBP plot acl
to IN go prep
a DT party det
church NN party compound
party NN to pobj
, , go punct
drink VBP go conj
and CC drink cc
then RB drive advmod
drive VB drink conj
. . go punct
```

The preceding code uses spaCy to parse the first sentence in the movie review corpus. After the parse, we iterate through the tokens in the resulting document and print the token and its part of speech tag. These are followed by the text of the token's head, or the word that it depends on, and the kind of dependency between the head and the token. For example, the word `couples` is tagged as a plural noun, it is dependent on the word `go`, and the dependency is `nsubj` since `couples` is the noun subject (`nsubj`) of `go`. In *Chapter 4*, we saw an example of a visualization of a dependency parse (*Figure 4.6*), which represents the dependencies as arcs between the item and its head; however, in the preceding code, we see more of the underlying information.

In this section, we have seen some examples of symbolic representations of language based on analyzing individual words and phrases, including parts of speech of individual words and labels for phrases. We can also represent words and phrases using a completely different and completely numeric approach based on vectors.

Representing language numerically with vectors

A common mathematical technique for representing language in preparation for machine learning is through the use of vectors. Both documents and words can be represented with vectors. We'll start by discussing document vectors.

Understanding vectors for document representation

We have seen that texts can be represented as sequences of symbols such as words, which is the way that we read them. However, it is usually more convenient for computational NLP purposes to represent text numerically, especially if we are dealing with large quantities of text. Another advantage of numerical representation is that we can also process text represented numerically with a much wider range of mathematical techniques.

A common way to represent both documents and words is by using vectors, which are basically one-dimensional arrays. Along with words, we can also use vectors to represent other linguistic units, such as lemmas or stemmed words, which were described in *Chapter 5*.

Binary bag of words

In *Chapter 3* and *Chapter 6*, we briefly discussed the **bag of words** (**BoW**) approach, where each document in a corpus is represented by a vector whose length is the size of the vocabulary. Each position in the vector corresponds to a word in the vocabulary and the element in the vector is 1 or 0, depending on whether that word occurs in that document. Each position in the vector is a feature of the document – that is, whether or not the word occurs. This is the simplest form of the BoW, and it is called the **binary bag of words**. It is immediately clear that this is a very coarse way of representing documents. All it cares about is whether a word occurs in a document, so it fails to capture a lot of information – what words are nearby, where in the document the words occur, and how often the words occur are all missing in the binary BoW. It is also affected by the lengths of documents since longer documents will have more words.

A more detailed version of the BoW approach is to count not just whether a word appears in a document but also how many times it appears. For this, we'll move on to the next technique, **count bag of words**.

Count bag of words

It seems intuitive that the number of times a word occurs in a document would help us decide how similar two documents are to each other. However, so far, we haven't used that information. In the document vectors we've seen so far, the values are just one and zero – one if the word occurs in the document and zero if it doesn't. If instead, we let the values represent the number of times the word occurs in the document, then we have more information. The BoW that includes the frequencies of the words in a document is a **count BoW**.

We saw the code to generate the binary BoW in the *Bag of words and k-means clustering* section (*Figure 6.15*) in *Chapter 6*. The code can be very slightly modified to compute a count BoW. The only change that has to be made is to increment the total count for a word when it is found more than once in a document. This is shown in the following code:

```
def document_features(document):
    features = {}
    for word in word_features:
        features[word] = 0
        for doc_word in document:
            if word == doc_word:
                features[word] += 1
    return features
```

Comparing this code to the code in *Figure 6.15*, we can see that the only difference is that the value of features[word] is incremented when the word is found in the document, rather than set to 1. The resulting matrix, shown in *Figure 7.2*, has many different values for word frequencies than the matrix in *Figure 6.16*, which had only zeros and ones:

	film	one	movie	like	even	good	time	story	would	much	...	spielberg	development	etc	language	blue	proves	vampire	seemingly	basic	caught
0	3	2	0	1	1	0	3	0	2	4	...	0	0	0	0	0	0	0	0	1	0
1	6	0	3	0	1	3	0	0	1	2	...	0	0	0	0	0	0	0	0	1	0
2	1	0	1	0	0	1	1	2	0	0	...	0	0	0	0	0	0	0	1	0	1
3	9	3	0	1	0	1	1	1	1	0	...	0	0	0	0	0	0	0	0	1	0
4	5	3	1	0	1	1	1	0	0	0	...	0	0	0	0	0	0	0	0	0	0
5	0	0	1	2	0	0	0	0	0	0	...	0	0	0	0	0	0	0	0	0	0
6	0	4	3	1	0	0	0	0	0	0	...	0	0	0	0	0	0	0	0	0	0
7	5	2	1	3	1	1	1	1	1	1	...	0	0	0	0	0	0	0	0	0	0
8	4	4	3	0	1	2	3	1	0	1	...	0	0	0	0	0	0	0	0	0	0
9	4	1	1	1	0	0	1	1	2	6	...	0	0	0	0	0	0	0	0	0	0

10 rows × 1000 columns

Figure 7.2 – Count BoW for the movie review corpus

In *Figure 7.2*, we're looking at 10 randomly selected documents, numbered from 0 to 9 in the first column. Looking more closely at the frequencies of film, (recall that film is the most common non-stopword in the corpus), we can see that all of the documents except document 5 and document 6 have at least one occurrence of film. In a binary BoW, they would all be lumped together, but here they have different values, which allows us to make finer-grained distinctions among documents. At this point, you might be interested in going back to the clustering exercise in *Chapter 6*, modifying the code in *Figure 6.22* to use the count BoW, and looking at the resulting clusters.

We can see from *Figure 7.2* that the count BoW gives us some more information about the words that occur in the documents than the binary BoW. However, we can do an even more precise analysis by using a technique called **term frequency-inverse document frequency (TF-IDF)**, which we describe in the next section.

Term frequency-inverse document frequency

Considering that our goal is to find a representation that accurately reflects similarities between documents, we can make use of some other insights. Specifically, consider the following observations:

- The raw frequency of words in a document will vary based on the length of the document. This means that a shorter document with fewer overall words might not appear to be similar to a longer document that has more words. So, we should be considering the proportion of the words in the document rather than the raw number.

- Words that occur very often overall will not be useful in distinguishing documents, because every document will have a lot of them. Clearly, the most problematic words that commonly occur within a corpus will be the stopwords we discussed in *Chapter 5*, but other words that are not strictly stopwords can have this property as well. Recall when we looked at the movie review corpus in *Chapter 6* that the words film and movie were very common in both positive and negative reviews, so they won't be able to help us tell those categories apart. The most helpful words will probably be the words that occur with different frequencies in different categories.

A popular way of taking these concerns into account is the measure *TF-IDF*. TF-IDF consists of two measurements – **term frequency (TF)** and **inverse document frequency (IDF)**.

TF is the number of times a term (or word) appears in a document, divided by the total number of terms in the document (which takes into account the fact that longer documents have more words overall). We can define that value as tf(term, document). For example, in *Figure 7.2*, we can see that the term film appeared once in document 0, so tf("film",0) is 1 over the length of document 0. Since the second document contains film four times, tf("film",1) is 4 over the length of document 1. The formula for term frequency is as follows:

$$\text{tf}(t,d) = \frac{f_{t,d}}{\sum_{t' \in d} f_{t',d}}$$

However, as we saw with stopwords, very frequent words don't give us much information for distinguishing documents. Even if TF(term, document) is very large, that could just be due to the fact that term occurs frequently in every document. To take care of this, we introduce IDF. The numerator of idf(term, Documents) is the total number of documents in the corpus, N, which we divide by the number of documents (D) that contain the term, t. In case the term doesn't appear in the corpus, the denominator would be 0, so 1 is added to the denominator to prevent division by 0. idf(term, documents) is the log of this quotient. The formula for idf is as follows:

$$idf(t, D) \; = \; \log \frac{N}{|\{d \in D : t \in d\}|}$$

Then the *TF-IDF* value for a term in a document in a given corpus is just the product of its TF and IDF:

$$tfidf(t, d, D) \; = \; tf \cdot idf(t, D)$$

To compute the TF-IDF vectors of the movie review corpus, we will use another very useful package, called scikit-learn, since NLTK and spaCy don't have built-in functions for TF-IDF. The code to compute TF-IDF in those packages could be written by hand using the standard formulas we saw in the preceding three equations; however, it will be faster to implement if we use the functions in scikit-learn, in particular, tfidfVectorizer in the feature extraction package. The code to compute TF-IDF vectors for the movie review corpus of 2,000 documents is shown in *Figure 7.3*. In this example, we will look only at the top 200 terms.

A tokenizer is defined in lines 9-11. In this example, we are using just the standard NLTK tokenizer, but any other text-processing function could be used here as well. For example, we might want to try using stemmed or lemmatized tokens, and these functions could be included in the tokenize() function. Why might stemming or lemmatizing text be a good idea?

One reason that this kind of preprocessing could be useful is that it will reduce the number of unique tokens in the data. This is because words that have several different variants will be collapsed into their root word (for example, *walk*, *walks*, *walking*, and *walked* will all be treated as the same word). If we believe that this variation is mostly just a source of noise in the data, then it's a good idea to collapse the variants by stemming or lemmatization. However, if we believe that the variation is important, then it won't be a good idea to collapse the variants, because this will cause us to lose information. We can make this kind of decision a priori by thinking about what information is needed for the goals of the application, or we can treat this decision as a hyperparameter, exploring different options and seeing how they affect the accuracy of the final result.

Figure 7.3 shows a screenshot of the code.

```python
1  import string
2  import os
3  from sklearn.feature_extraction.text import TfidfVectorizer
4
5  # consider only the most common words
6  max_tokens = 200
7
8  #this tokenizer will be used to tokenize the inputs
9  def tokenize(text):
10     tokens = nltk.word_tokenize(text)
11     return tokens
12
13 path = './movie_reviews/'
14 token_dict = {}
15
16 # look at all the files in the given path, there are 2,000 files
17 for dirpath, dirs, files in os.walk(path):
18     for f in files:
19         fname = os.path.join(dirpath, f)
20         with open(fname) as review:
21             text = review.read()
22             token_dict[f] = text.lower().translate(str.maketrans('', '',string.punctuation))
23
24 #get a new tfIdf vectorizer
25 tfIdfVectorizer = TfidfVectorizer(input = "content",
26                                   use_idf = True,
27                                   tokenizer = tokenize,
28                                   max_features = max_tokens,
29                                   stop_words = 'english')
30
31 # use the vectorizer to compute the tfIdf of the dataset
32 tfIdf = tfIdfVectorizer.fit_transform(token_dict.values())
33
34 # the feature names are the words (tokens) in the dataset
35 tfidf_tokens = tfIdfVectorizer.get_feature_names_out()
36
37 final_vectors = pd.DataFrame(
38     data = tfIdf.toarray(),
39     columns = tfidf_tokens
40 )
41
42 final_vectors
```

Figure 7.3 – Code to compute TF-IDF vectors for the movie review corpus

Returning to *Figure 7.3*, the next step after defining the tokenization function is to define the path where the data can be found (line 13) and initialize the token dictionary at line 14. The code then walks the data directory and collects the tokens from each file. While the code is collecting tokens, it lowercases the text and removes punctuation (line 22). The decisions to lowercase the text and remove punctuation are up to the developer, similar to the decision we discussed previously on whether or not to stem or lemmatize the tokens. We could explore empirically whether or not these two preprocessing steps improve processing accuracy; however, if we think about how much meaning is carried by case and punctuation, it seems clear that in many applications, case and punctuation don't add much meaning. In those kinds of applications, lowercasing the text and removing punctuation will improve the results.

Having collected and counted the tokens in the full set of files, the next step is to initialize tfIdfVectorizer, which is a built-in function in scikit-learn. This is accomplished in lines 25-29. The parameters include the type of input, whether or not to use IDF, which tokenizer to use, how many features to use, and the language for stopwords (English, in this example).

Line 32 is where the real work of TF-IDF is done, with the fit_transform method, where the TF-IDF vector is built from the documents and the tokens. The remaining code (lines 37-42) is primarily to assist in displaying the resulting TF-IDF vectors.

The resulting TF-IDF matrix is shown in *Figure 7.4*:

	2	acting	action	actor	actors	actually	alien	american	audience
0	0.092834	0.000000	0.000000	0.000000	0.069898	0.134576	0.0	0.084837	0.130063
1	0.000000	0.266255	0.132484	0.000000	0.000000	0.000000	0.0	0.000000	0.000000
2	0.000000	0.094612	0.000000	0.000000	0.000000	0.000000	0.0	0.115113	0.264719
3	0.000000	0.000000	0.000000	0.000000	0.160810	0.000000	0.0	0.000000	0.000000
4	0.000000	0.000000	0.000000	0.210719	0.000000	0.090622	0.0	0.000000	0.087584
...	...	...	...	...	...	...	...	...	...
1995	0.000000	0.046388	0.000000	0.000000	0.000000	0.000000	0.0	0.056440	0.000000
1996	0.000000	0.000000	0.000000	0.171716	0.000000	0.147697	0.0	0.000000	0.000000
1997	0.000000	0.000000	0.133597	0.000000	0.000000	0.000000	0.0	0.000000	0.000000
1998	0.000000	0.000000	0.000000	0.089530	0.159988	0.000000	0.0	0.000000	0.000000
1999	0.000000	0.000000	0.000000	0.000000	0.000000	0.000000	0.0	0.000000	0.000000

2000 rows × 200 columns

Figure 7.4 – Partial TF-IDF vectors for some of the documents in the movie review corpus

We have now represented our movie review corpus as a matrix where every document is a vector of N dimensions, where N is the maximum size of vocabulary we are going to use. *Figure 7.4* shows the TF-IDF vectors for a few of the 2,000 documents in the corpus (documents 0-4 and 1995-1999), shown in the rows, and some of the words in the corpus, shown in alphabetical order at the top. Both the words and the documents are truncated for display purposes.

In *Figure 7.4*, we can see that there is quite a bit of difference in the TF-IDF values for the same words in different documents. For example, `acting` has considerably different scores in documents 0 and 1. This will become useful in the next step of processing (classification), which we will return to in *Chapter 9*. Note that we have not done any actual machine learning yet; so far, the goal has been simply to convert documents to numerical representations, based on the words they contain.

Up until now, we've been focusing on representing documents. But what about representing the words themselves? The words in a document vector are just numbers representing their frequency, either just in a document or their frequency in a document relative to their frequency in a corpus (TF-IDF). We don't have any information about the meanings of the words themselves in the techniques we've looked at so far. However, it seems clear that the meanings of words in a document should also impact the document's similarity to other documents. We will look at representing the meanings of words in the next section. This representation is often called **word embeddings** in the NLP literature. We will begin with a popular representation of words as vectors, **Word2Vec**, which captures the similarity in meaning of words to each other.

Representing words with context-independent vectors

So far, we have looked at several ways of representing similarities among documents. However, finding out that two or more documents are similar to each other is not very specific, although it can be useful for some applications, such as intent or document classification. In this section, we will talk about representing the meanings of words with word vectors.

Word2Vec

Word2Vec is a popular library for representing words as vectors, published by Google in 2013 (Mikolov, Tomas; et al. (2013). *Efficient Estimation of Word Representations in Vector Space*. https://arxiv.org/abs/1301.3781). The basic idea behind Word2Vec is that every word in a corpus is represented by a single vector that is computed based on all the contexts (nearby words) in which the word occurs. The intuition behind this approach is that words with similar meanings will occur in similar contexts. This intuition is summarized in a famous quote from the linguist J. R. Firth, "*You shall know a word by the company it keeps*" (*Studies in Linguistic Analysis*, Wiley-Blackwell).

Let's build up to Word2Vec by starting with the idea of assigning each word to a vector. The simplest vector that we can use to represent words is the idea of **one-hot encoding**. In one-hot encoding, each word in the vocabulary is represented by a vector where that word has 1 in a specific position in the vector, and all the rest of the positions are zeros (it is called one-hot encoding because one bit is on

– that is, *hot*). The length of the vector is the size of the vocabulary. The set of one-hot vectors for the words in a corpus is something like a dictionary in that, for example, we could say that if the word is `movie`, it will be represented by 1 in a specific position. If the word is `actor`, it will be represented by 1 in a different position.

At this point, we aren't taking into account the surrounding words yet. The first step in one-hot encoding is integer encoding, where we assign a specific integer to each word in the corpus. The following code uses libraries from scikit-learn to do the integer encoding and the one-hot encoding. We also import some functions from the `numpy` library, `array` and `argmax`, which we'll be returning to in later chapters:

```
from numpy import array
from numpy import argmax
from sklearn.preprocessing import LabelEncoder
from sklearn.preprocessing import OneHotEncoder

#import the movie reviews
from nltk.corpus import movie_reviews

# make a list of movie review documents
documents = [(list(movie_reviews.words(fileid)))
             for category in movie_reviews.categories()
             for fileid in movie_reviews.fileids(category)]

# for this example, we'll just look at the first document, and
# the first 50 words
data = documents[0]
values = array(data)
short_values = (values[:50])

# first encode words as integers
# every word in the vocabulary gets a unique number
label_encoder = LabelEncoder()
integer_encoded = label_encoder.fit_transform(short_values)

# look at the first 50 encodings
print(integer_encoded)
[32  3 40 35 12 19 39  5 10 31  1 15  8 37 16  2 38 17 26  7  6  2 30
29
 36 20 14  1  9 24 18 11 39 34 23 25 22 27  1  8 21 28  2 42  0 33 36
13
  4 41]
```

Looking at the first 50 integer encodings in the first movie review, we see a vector of length 50. This can be converted to a one-hot encoding, as shown in the following code:

```
# convert the integer encoding to onehot encoding
onehot_encoder = OneHotEncoder(sparse=False)
integer_encoded = integer_encoded.reshape(
    len(integer_encoded), 1)
onehot_encoded = onehot_encoder.fit_transform(
    integer_encoded)

print(onehot_encoded)
# invert the first vector so that we can see the original word it
encodes
inverted = label_encoder.inverse_transform(
    [argmax(onehot_encoded[0, :])])
print(inverted)
[[0. 0. 0. ... 0. 0. 0.]
 [0. 0. 0. ... 0. 0. 0.]
 [0. 0. 0. ... 1. 0. 0.]
 ...
 [0. 0. 0. ... 0. 0. 0.]
 [0. 0. 0. ... 0. 0. 0.]
 [0. 0. 0. ... 0. 1. 0.]]
['plot']
```

The output in the preceding code shows first a subset of the one-hot vectors. Since they are one-hot vectors, they have 0 in all positions except one position, whose value is 1. Obviously, this is very sparse, and it would be good to provide a more condensed representation.

The next to the last line shows how we can invert a one-hot vector to recover the original word. The sparse representation takes a large amount of memory and is not very practical.

The Word2Vec approach uses a neural net to reduce the dimensionality of the embedding. We will be returning to the details of neural networks in *Chapter 10*, but for this example, we'll use a library called Gensim that will compute Word2Vec for us.

The following code uses the Gensim `Word2Vec` library to create a model of the movie review corpus. The Gensim `model` object that was created by Word2Vec includes a large number of interesting methods for working with the data. The following code shows one of these – `most_similar`, which, given a word, can find the words that are the most similar to that word in the dataset. Here, we can see a list of the 25 words most similar to `movie` in the corpus, along with a score that indicates how similar that word is to `movie`, according to the Word2Vec analysis:

```
import gensim
import nltk

from nltk.corpus import movie_reviews
from gensim.models import Word2Vec

# make a list of movie review documents
documents = [(list(movie_reviews.words(fileid)))
                for category in movie_reviews.categories()
                for fileid in movie_reviews.fileids(category)]
all_words = movie_reviews.words()
model = Word2Vec(documents, min_count=5)
model.wv.most_similar(positive = ['movie'],topn = 25)

[('film', 0.9275647401809692),
 ('picture', 0.8604983687400818),
 ('sequel', 0.7637531757354736),
 ('flick', 0.7089548110961914),
 ('ending', 0.6734793186187744),
 ('thing', 0.6730892658233643),
 ('experience', 0.6683703064918518),
 ('premise', 0.6510635018348694),
 ('comedy', 0.6485130786895752),
 ('genre', 0.6462267637252808),
 ('case', 0.6455731391906738),
 ('it', 0.6344209313392639),
 ('story', 0.6279274821281433),
 ('mess', 0.6165297627449036),
 ('plot', 0.6162343621253967),
 ('message', 0.6131927371025085),
 ('word', 0.6131172776222229),
 ('movies', 0.6125075221061707),
 ('entertainment', 0.6109789609909058),
 ('trailer', 0.6068858504295349),
 ('script', 0.6000528335571289),
```

```
('audience', 0.5993804931640625),
('idea', 0.5915037989616394),
('watching', 0.5902948379516602),
('review', 0.581495584487915)]
```

As can be seen in the preceding code, the words that Word2Vec finds to be the most similar to `movie` based on the contexts in which they occur are very much what we would expect. The top two words, `film` and `picture`, are very close synonyms of `movie`. In *Chapter 10*, we will return to Word2Vec and see how this kind of model can be used in NLP tasks.

While Word2Vec does take into account the contexts in which words occur in a dataset, every word in the vocabulary is represented by a single vector that encapsulates all of the contexts in which it occurs. This glosses over the fact that words can have different meanings in different contexts. The next section reviews approaches representing words depending on specific contexts.

Representing words with context-dependent vectors

Word2Vec's word vectors are context-independent in that a word always has the same vector no matter what context it occurs in. However, in fact, the meanings of words are strongly affected by nearby words. For example, the meanings of the word *film* in *We enjoyed the film* and *the table was covered with a thin film of dust* are quite different. To capture these contextual differences in meanings, we would like to have a way to have different vector representations of these words that reflect the differences in meanings that result from the different contexts. This research direction has been extensively explored in the last few years, starting with the **BERT (Bidirectional Encoder Representations from Transformers)** system (`https://aclanthology.org/N19-1423/` (Devlin et al., NAACL 2019)).

This approach has resulted in great improvements in NLP technology, which we will want to discuss in depth. For that reason, we will postpone a fuller look at context-dependent word representations until we get to *Chapter 11*, where we will address this topic in detail.

Summary

In this chapter, we've learned how to select different NLP approaches, based on the available data and other requirements. In addition, we've learned about representing data for NLP applications. We've placed particular emphasis on vector representations, including vector representations of both documents and words. For documents, we've covered binary bag of words, count bag of words, and TF-IDF. For representing words, we've reviewed the Word2Vec approach and briefly introduced context-dependent vectors, which will be covered in much more detail in *Chapter 11*.

In the next four chapters, we will take the representations that we've learned about in this chapter and show how to train models from them that can be applied to different problems such as document classification and intent recognition. We will start with rule-based techniques in *Chapter 8*, discuss traditional machine learning techniques in *Chapter 9*, talk about neural networks in *Chapter 10*, and discuss the most modern approaches, transformers, and pretrained models in *Chapter 11*.

8

Rule-Based Techniques

Rule-based techniques are a very important and useful tool in **natural language processing** (NLP). Rules are used to examine text and decide how it should be analyzed in an all-or-none fashion, as opposed to the statistical techniques we will be reviewing in later chapters. In this chapter, we will discuss how to apply rule-based techniques to NLP. We will look at examples such as regular expressions, syntactic parsing, and semantic role assignment. We will primarily use the NLTK and spaCy libraries, which we have seen in earlier chapters.

In this chapter, we will cover the following topics:

- Rule-based techniques

- Why use rules?

- Exploring regular expressions

- Sentence-level analysis – syntactic parsing and semantic role assignment

Rule-based techniques

Rule-based techniques in NLP are, as the name suggests, based on rules written by human developers, as opposed to machine-learned models derived from data. Rule-based techniques were, for many years, the most common approach to NLP, but as we saw in *Chapter 7*, rule-based approaches have largely been superseded by numerical, machine-learned approaches for the overall design of most NLP applications. There are many reasons for this; for example, since rules are written by humans, it is possible that they might not cover all situations if the human developer has overlooked something.

However, for practical applications, rules can be very useful, either by themselves or, more likely, along with machine-learned models.

The next section will discuss the motivations for using rules in NLP applications.

Why use rules?

Rules are a useful approach in one or more of the following situations:

- The application you are developing has a requirement to analyze fixed expressions that include thousands or even millions of variants when it would be extremely difficult to provide enough data to learn a machine model. These kinds of expressions include numbers, monetary amounts, dates, and addresses, for example. It is hard for systems to learn models when the data is so diverse. Moreover, it is usually not necessary, as rules to analyze these expressions are not difficult to write because their formats are very structured. For both of these reasons, a rule-based approach is a simpler solution for recognizing fixed expressions.

- Very little training data is available for the application, and creating new data would be expensive. For example, annotating new data might require very specialized expertise. Although there are now techniques (such as few-shot or zero-shot learning) that can adapt large pre-trained models to cover specialized domains, if the domain-specific data is very different from the original training data in terms of syntax or vocabulary, it will be difficult for the adaptation process to work well. Medical reports and air traffic control messages are examples of these kinds of data.

- There are existing, well-tested rules and libraries available that can easily be used in new applications, such as the Python `datetime` package for recognizing dates and times.

- The goal is to bootstrap a machine-learned model by preliminary annotation of a corpus. Corpus annotation is needed in preparation for using that corpus as training data for a machine learning model, or for using the corpus as a *gold standard* in NLP system evaluation. In this process, the data is first annotated by applying some hand-written rules. Then, the resulting annotated corpus is usually reviewed and corrected by human annotators, since it is likely to contain errors. Even though the corpus requires a review process, an initial annotation by a rule-based system will save time over manual annotation from scratch.

- The application needs to find named entities from a fixed, known set.

- The results have to be very precise – for example, grammar checking, proofreading, language learning, and authorship studies.

- A quick prototype is needed in order to test downstream processing, and the data collection and model training stages needed for machine learning would take too much time.

We will start by looking at regular expressions, a very common technique for analyzing text that contains well-understood patterns.

Exploring regular expressions

Regular expressions are a widely used rule-based technique that is often used for recognizing fixed expressions. By **fixed expressions**, we mean words and phrases that are formed according to their own internal rules, which are largely different from the normal patterns of the language.

One type of fixed expression is *monetary amounts*. There are only a few variations in formats for monetary amounts – the number of decimal places, the symbol for the type of currency, and whether the numbers are separated by commas or periods. The application might only have a requirement to recognize specific currencies, which would simplify the rules further. Other common fixed expressions include *dates*, *times*, *telephone numbers*, *addresses*, *email addresses*, *measurements*, and *numbers*. Regular expressions in NLP are most frequently used in preprocessing text that will be further analyzed with other techniques.

Different programming languages have slightly different formats for regular expressions. We will be using the Python formats defined at `https://docs.python.org/3/library/re.html` and available in the Python `re` library. We will not define regular expression syntax here because there are many resources on the web that describe regular expression syntax, including the Python documentation, and we don't need to duplicate that. You might find the information at `https://www.h2kinfosys.com/blog/nltk-regular-expressions/` and `https://python.gotrained.com/nltk-regex/` useful for getting into the details of regular expressions in NLTK.

We will start by going over the basics of operating on strings with regular expressions, followed by some tips for making it easier to apply and debug regular expressions.

Recognizing, parsing, and replacing strings with regular expressions

The simplest use of regular expressions is to simply note that a match occurred. What we want to do after a fixed expression has been matched depends on the goal of the application. In some applications, all we want to do is recognize that a fixed expression has occurred or did not occur. This can be useful, for example, in validating user input in web forms, so that users can correct invalid address formats. The following code shows how to recognize a US address with regular expressions:

```
import re
# process US street address
# the address to match
text = "223 5th Street NW, Plymouth, PA 19001"
print(text)
# first define components of an address
# at the beginning of a string, match at least one digit
street_number_re = "^\d{1,}"
```

```
# match street names containing upper and lower case letters and
digits, including spaces,
# followed by an optional comma
street_name_re = "[a-zA-Z0-9\s]+,?"


# match city names containing letters, but not spaces, followed by a
comma
# note that two word city names (like "New York") won't get matched
# try to modify the regular expression to include two word city names
city_name_re = " [a-zA-Z]+(\,)?"


# to match US state abbreviations, match any two upper case alphabetic
characters
# notice that this overgenerates and accepts state names that don't
exist because it doesn't check for a valid state name
state_abbrev_re = " [A-Z]{2}"


# match US postal codes consisting of exactly 5 digits. 9 digit codes
exist, but this expression doesn't match them
postal_code_re = " [0-9]{5}$"
# put the components together -- define the overall pattern
address_pattern_re = street_number_re + street_name_re + city_name_re
+ state_abbrev_re + postal_code_re


# is this an address?
is_match = re.match(address_pattern_re,text)
if is_match is not None:
    print("matches address_pattern")
else:
    print("doesn't match")
```

In other cases, we might want to parse the expression and assign meanings to the components – for example, in a date, it could be useful to recognize a month, a day, and a year. In still other cases, we might want to replace the expression with another expression, delete it, or normalize it so that all occurrences of the expression are in the same form. We could even want to do a combination of these operations. For example, if the application is classification, it is likely that we only need to know whether or not the regular expression occurred; that is, we don't need its content. In that case, we can replace the expression with a class token such as DATE, so that a sentence such as *We received the package on August 2, 2022* becomes *We received the package on DATE*. Replacing the whole expression with a class token can also be used for redacting sensitive text such as social security numbers.

The preceding code shows how to use regular expressions to match patterns in text and shows the code to confirm a match. However, this example just shows how to confirm that a match exists. We might want to do other things, such as replace the address with a class label, or label the matched portion of a string. The following code shows how to use the regular expression sub method to substitute a class label for an address:

```
# the address to match
text = "223 5th Street NW, Plymouth, PA 19001"
# replace the whole expression with a class tag -- "ADDRESS"
address_class = re.sub(address_pattern_re,"ADDRESS",text)
print(address_class)
ADDRESS
```

Another useful operation is to label the whole expression with a semantic label such as address, as shown in the following code. This code shows how to add a label to the address. This enables us to identify US addresses in text and do tasks such as counting them or extracting all the addresses from texts:

```
# suppose we need to label a matched portion of the string
# this function will label the matched string as an address
def add_address_label(address_obj):
    labeled_address = add_label("address",address_obj)
    return(labeled_address)
# this creates the desired format for the labeled output
def add_label(label, match_obj):
    labeled_result = "{" + label + ":" + "'" + match_obj.group() + "'"
+ "}"
    return(labeled_result)

# add labels to the string
address_label_result = re.sub(address_pattern_re,add_address_
label,text)
print(address_label_result)
```

The result of running the preceding code is as follows:

```
{address:'223 5th Street NW, Plymouth, PA 19001'}
```

Finally, regular expressions are useful if we want to remove the whole match from the text – for example, to remove HTML markup.

General tips for using regular expressions

Regular expressions can easily become very complex and difficult to modify and debug. They can also easily fail to recognize some examples of what they're supposed to recognize and falsely recognize what they're not supposed to recognize. While it is tempting to try to match the regular expression so that it recognizes exactly what it is supposed to recognize and nothing else, this can make the expression so complicated that it is difficult to understand. Sometimes, it can be better to miss a few edge cases to keep the expression simple.

If we find that an existing regular expression is failing to find some expressions that we want to capture, or incorrectly finding expressions that we don't want to capture, it can sometimes be difficult to revise the existing expression without breaking some things that used to work. Here are a few tips that can make regular expressions easier to work with:

- Write down what you want the regular expression to match first (such as *any two consecutive uppercase alphabetic characters*). This will be helpful in both clarifying what you're trying to do as well as in helping catch any cases that you might have overlooked.

- Break complex expressions into components and test each component independently before putting them together. Besides helping with debugging, the component expressions can potentially be reused in other complex expressions. We saw this in the first code block in the previous section with components such as `street_name_re`.

- Use existing tested regular expressions for common expressions, for example, the Python `datetime` package (see `https://docs.python.org/3/library/datetime.html`), before trying to write your own regular expressions. They have been well tested over many years by many developers.

The next two sections cover specific ways to analyze two of the most important aspects of natural language: words and sentences.

The next section will start this topic by talking about analyzing individual words.

Word-level analysis

This section will discuss two approaches to analyzing words. The first one, lemmatization, involves breaking words down into their components in order reduce the variation in texts. The second one discusses some ideas for making use of hierarchically organized semantic information about the meanings of words in the form of ontologies.

Lemmatization

In our earlier discussion of preprocessing text in *Chapter 5*, we went over the task of **lemmatization** (and the related task of stemming) as a tool for regularizing text documents so that there is less variation in the documents we are analyzing. As we discussed, the process of lemmatization converts each word in the text to its root word, discarding information such as plural endings like *-s* in English. Lemmatization also requires a dictionary, because the dictionary supplies the root words for the words being lemmatized. We used Princeton University's **WordNet** (`https://wordnet.princeton.edu/`) as a dictionary when we covered lemmatization in *Chapter 5*.

We will use WordNet's semantic information about the relationships among words in the next section, where we discuss ontologies and their applications.

Ontologies

Ontologies, which we briefly mentioned in *Chapter 3* in the *Semantic analysis* section, represent the meanings of related words in hierarchical structures like the one shown in *Figure 3.2* for the word *airplane*. The ontology in *Figure 3.2* represents the information that an airplane is a type of heavier-than-aircraft, which is a type of craft, which is a type of vehicle, and so on up to the very generic, top-level category, `entity`.

The ontology in *Figure 3.2* is part of the **WordNet** ontology for English and a number of other languages. These hierarchical relationships are sometimes called **is a** relationships. For example, *an airplane is a vehicle*. In this example, we say that *vehicle* is a **superordinate** term and *airplane* is a **subordinate** term. WordNet uses some of its own terminology. In WordNet terminology, **hypernym** is the same thing as a superordinate term, and **hyponym** is the same thing as a subordinate term.

WordNet also includes many other semantic relationships, such as synonymy and *part-whole*. For example, we can find out that a wing is part of an airplane from WordNet. In addition, WordNet also includes part-of-speech information, and you will recall that we used this part-of-speech information in *Chapter 5* for part-of-speech tagging texts in preparation for lemmatization.

There are other ontologies besides WordNet, and you can even construct your own ontologies with tools such as Stanford University's Protégé (`https://protege.stanford.edu/`). However, WordNet is a good way to get started.

How can we make use of ontologies such as WordNet in NLP applications? Here are a few ideas:

- Develop a writing tool that helps authors find synonyms, antonyms, and definitions of words they would like to use.

- Count the number of mentions of different categories of words in texts. For example, you might be interested in finding all mentions of vehicles. Even if the text actually says *car* or *boat*, you could tell that the text mentions a vehicle by looking for words with *vehicle* as their hypernym.

- Generate additional examples of training data for machine learning by substituting different words with the same superordinate term in different sentence patterns. For example, suppose we have a chatbot that provides advice about cooking. It would probably get questions such as *How can I tell whether a pepper is ripe?*, or *Can I freeze tomatoes?* There are hundreds of types of food that could be substituted for *pepper* and *tomatoes* in those questions. It would be very tedious to create training examples for all of them. To avoid this, you could find all of the different types of vegetables in WordNet and generate training data by putting them into sentence templates to create new sentences.

Let's see an example of the previous strategy.

You probably recall from earlier mentions of WordNet that it is included in NLTK, so we can import it and ask for the list of senses of *vegetable* (**synsets**) as follows:

```
import nltk
from nltk.corpus import wordnet as wn
wn.synsets('vegetable')
```

We'll then see that there are two *senses*, or meanings, of *vegetable*, and we can ask for their definitions in the following code:

```
[Synset('vegetable.n.01'), Synset('vegetable.n.02')]
print(wn.synset('vegetable.n.01').definition())
print(wn.synset('vegetable.n.02').definition())
```

The format of the sense names such as `vegetable.n.01` should be interpreted as `word` and `part-of-speech` (*n* means *noun* here), followed by the word's order in the list of senses. We print the definitions of each of the two senses so that we can see what the WordNet senses mean. The resulting definitions are as follows:

```
edible seeds or roots or stems or leaves or bulbs or tubers or
nonsweet fruits of any of numerous herbaceous plant
any of various herbaceous plants cultivated for an edible part such as
the fruit or the root of the beet or the leaf of spinach or the seeds
of bean plants or the flower buds of broccoli or cauliflower
```

The first sense refers to the part that we eat, and the second sense refers to the plants whose parts we eat. If we are interested in cooking, we probably want the first sense of *vegetable* as *food*. Let's get the list of all the vegetables in the first sense, using the following code:

```
word_list = wn.synset('vegetable.n.01').hyponyms()

simple_names = []
for word in range(len(word_list)):
    simple_name = word_list[word].lemma_names()[0]
    simple_names.append(simple_name)
print(simple_names)
['artichoke', 'artichoke_heart', 'asparagus', 'bamboo_shoot',
'cardoon', 'celery', 'cruciferous_vegetable', 'cucumber', 'fennel',
'greens', 'gumbo', 'julienne', 'leek', 'legume', 'mushroom', 'onion',
'pieplant', 'plantain', 'potherb', 'pumpkin', 'raw_vegetable', 'root_
vegetable', 'solanaceous_vegetable', 'squash', 'truffle']
```

The code goes through the following steps:

1. Collect all the different types of the first sense of *vegetable* (the hyponyms) and store them in the `word_list` variable.

2. Iterate through the list of words, collect the lemma for each word, and store the lemmas in the `simple_names` variable.

3. Print the words.

We can then generate some sample data by filling in a text template with each word, as follows:

```
text_frame = "can you give me some good recipes for "
for vegetable in range(len(simple_names)):
    print(text_frame + simple_names[vegetable])

can you give me some good recipes for artichoke
can you give me some good recipes for artichoke_heart
can you give me some good recipes for asparagus
can you give me some good recipes for bamboo_shoot
```

The preceding code shows the first few sentences that we can generate from the text frame and the list of vegetables. Of course, in a real application, we would want to have multiple text frames to get a good variety of sentences.

At the beginning of this section, we listed a few ways to apply ontologies in NLP applications; you can probably come up with more if you think of different ways that you could make use of the meanings of words to solve problems in natural language applications.

However, words don't occur in isolation; they are combined with other words to create sentences with richer and more complex meanings. The next section will move on from analyzing words to the analysis of entire sentences, which we will analyze both syntactically and semantically.

Sentence-level analysis

Sentences can be analyzed in terms of their **syntax** (the structural relationships among parts of the sentence) or their **semantics** (the relationships among the meanings of the parts of the sentence). We'll look at both of these types of analysis next. Recognizing syntactic relationships is useful on its own for applications such as grammar checking (does the subject of the sentence agree with the verb? Is the correct form of the verb being used?), while recognizing semantic relationships on their own is useful for applications such as finding the components of a request in chatbots. Recognizing both syntactic and semantic relationships together is an alternative to statistical methods in almost any NLP application.

Syntactic analysis

The syntax of sentences and phrases can be analyzed in a process called **parsing**. Parsing is a type of analysis that attempts to match a set of rules, called grammar, to an input text. There are many approaches to parsing. We will not cover this topic in detail since there are many other resources available if you are interested in this topic, such as `https://en.wikipedia.org/wiki/Syntactic_parsing_(computational_linguistics)` or `https://www.tutorialspoint.com/natural_language_processing/natural_language_processing_syntactic_analysis.htm`. Chart parsing, dependency parsing, and recursive descent parsing are only a few of the many approaches to syntactic parsing. NLTK includes a `parse` package, which includes a variety of parsing algorithms that you can explore. For our examples in this section, we will use the chart parser in the `nltk.parse.ChartParser` class, which is a common and basic approach.

Context-free grammars and parsing

A very common way to define the rules for syntactic parsing is **context-free grammars** (**CFGs**). CFGs can be used in chart parsing as well as many other parsing algorithms. You may be familiar with this format because it is widely used in computer science for defining formal languages, such as programming languages. CFGs consist of a set of *rules*. Each rule consists of a **left-hand side** (**LHS**) and a **right-hand side** (**RHS**), typically separated by a symbol, such as an arrow. The rule is interpreted to mean that the single symbol on the LHS is made up of the components of the RHS.

For example, the context-free rule `S -> NP VP` states that a sentence (`S`) consists of a noun phrase (**NP**), followed by a **verb phrase** (**VP**). An NP can consist of a **determiner** (**Det**) such as *an*, *my*, or *the*, followed by one or two **nouns** (**Ns**) such as *elephant*, possibly followed by a **prepositional phrase** (**PP**), or just a **pronoun** (**Pro**), and so on. Every rule must be in turn defined with another rule, until the rules end in words (or, more generally, *terminal symbols*), which do not appear on the LHS of any rule.

The following shows the code for creating a CFG for a few rules of English. These are *constituency rules*, which show how the parts of the sentence are related to each other. There is another commonly used format, *dependencies*, which shows how the words are related to each other, which we will not explore in this book because the constituency rules are sufficient to illustrate the basic concepts of syntactic grammar and syntactic parsing:

```
grammar = nltk.CFG.fromstring("""
S -> NP VP
PP -> P NP
NP -> Det N | Det N N |Det N PP | Pro
Pro -> 'I' |'you'|'we'
VP -> V NP | VP PP
Det -> 'an' | 'my' | 'the'
N -> 'elephant' | 'pajamas' | 'movie' |'family' | 'room' |'children'
V -> 'saw'|'watched'
P -> 'in'
""")
```

This grammar is only able to parse a few sentences, such as *the children watched the movie in the family room*. For example, it would not be able to parse a sentence *the children slept* because, in this grammar, the VP has to include an object or prepositional phrase in addition to the verb. A full CFG for English would be much larger and more complex than the one in the preceding code. It's also worth pointing out that the NLTK rules can be annotated with probabilities that indicate the likelihood of each alternative on the RHS.

For example, *rule 4* (Pro → 'I' | 'you' | 'we') in the preceding code could have probabilities for the relative likelihoods of *I*, *you*, and *we*. In practice, this will result in more accurate parses, but it does not affect the examples we'll show in this chapter. *Table 8.1* summarizes some CFG terminology:

Symbol	Meaning	Example
S	Sentence	The children watched the movie
NP	Noun phrase	The children
VP	Verb phrase	Watched the movie
PP	Prepositional phrase	In the family room
Pro	Pronoun	I, we, you, they, he, she, it
Det	Determiner or article	The, a
V	Verb	Watched, saw
N	Noun	Children, movie, elephant, family, room

Table 8.1 – Meanings of CFG terms for the grammar in the CFG code block

Table 8.2 summarizes some syntactic conventions used in NLTK CFGs:

Symbol	Meaning
->	Separates the LHS from the RHS
|	Separates alternate possibilities for RHSs that expand the LHS
Single quotes	Indicates a word; that is, a terminal symbol
Initial capitalization	Indicates a syntactic category; that is, a non-terminal symbol, which is expected to be defined by additional rules

Table 8.2 – CFG syntax

We can parse and visualize the sentence *the children watched the movie in the family room* with the grammar in the previous code block using the following code:

```
# we will need this to tokenize the input
from nltk import word_tokenize
```

```
# a package for visualizing parse trees
import svgling

# to use svgling we need to disable NLTK's normal visualization
functions
svgling.disable_nltk_png()

# example sentence that can be parsed with the grammar we've defined
sent = nltk.word_tokenize("the children watched the movie in the
family room")

# create a chart parser based on the grammar above
parser = nltk.ChartParser(grammar)

# parse the sentence
trees = list(parser.parse(sent))

# print a text-formatted parse tree
print(trees[0])

# print an SVG formatted parse tree
trees[0]
```

We can view the results of parsing in different ways – for example, as a bracketed text format, as in the following code:

```
(S
  (NP (Det the) (N children))
  (VP
    (VP (V watched) (NP (Det the) (N movie)))
    (PP (P in) (NP (Det the) (N family) (N room)))))))
```

Notice that the parse directly reflects the grammar: the overall result is called *S*, because it came from the first rule in the grammar, S -> NP VP. Similarly, *NP* and *VP* are connected directly to *S*, and their child nodes are listed in parentheses after them.

The preceding format is useful for possible later stages of processing that may need to be computer-readable; however, it is a little bit difficult to read. *Figure 8.1* shows a conventional tree diagram of this parse, which is easier to view. As in the case of the preceding text parse, you can see that it aligns directly with the grammar. The words, or terminal symbols, all appear at the bottom, or *leaves*, of the tree:

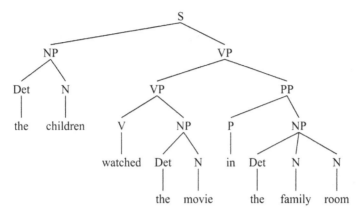

Figure 8.1 – Constituency tree for "the children watched the movie in the family room"

You can try parsing other sentences with this grammar, and you can also try modifying the grammar. For example, try adding a grammar rule that will enable the grammar to parse sentences with verbs that are not followed by NPs or PPs, such as *the children slept*.

Semantic analysis and slot filling

The previous sections on regular expressions and syntactic analysis dealt only with the structure of sentences, not their meaning. A syntactic grammar like the one shown in the preceding section can parse nonsense sentences such as *the movie watched the children in the room room* as long as they match the grammar. We can see this in *Figure 8.2*:

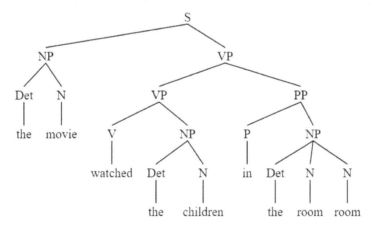

Figure 8.2 – Constituency tree for "the movie watched the children in the room room"

In most applications, however, we don't only want to find the syntactic structure of sentences; we also want to extract part or all of their meanings. The process of extracting meaning is called **semantic analysis**. The particular sense of *meaning* will vary depending on the application. For example, in many of the applications we've worked on so far in this book, the only meaning that needed to be derived from a document was its overall classification. This was the case in the movie review data – the only meaning that we wanted to get from the document was the positive or negative sentiment. The statistical methods that we've looked at in previous chapters are very good at doing this kind of coarse-grained processing.

However, there are other applications in which it's necessary to get more detailed information about the relationships between items in the sentence. While there are machine learning techniques for getting fine-grained information (and we will be discussing them in *Chapters 9, 10, and 11*), they work best with large amounts of data. If there is less data available, sometimes rule-based processing can be more effective.

Basic slot filling

For the next examples, we will be looking in detail at a technique often used in interactive applications, **slot filling**. This is a common technique used in voicebots and chatbots, although it is also used in non-interactive applications such as *information extraction*.

As an example, consider a chatbot application that helps a user find a restaurant. The application is designed to expect the user to offer some search criteria, such as the type of cuisine, the atmosphere, and the location. These criteria are the *slots* in the application. For example, the user might say, *I'd like to find a nice Italian restaurant near here*. The overall user goal, **Restaurant Search**, is the *intent*. We will be focusing on identifying slots in this chapter but will discuss intents in more detail in later chapters.

The design for this application is shown in *Figure 8.3*:

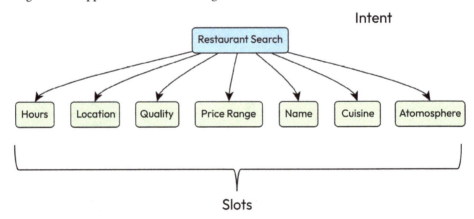

Figure 8.3 – Slots for a restaurant search application

In processing the user's utterance, the system has to identify what slots the user has specified, and it has to extract their values. This is all the information the system needs to help the user find a restaurant, so anything else in the sentence is normally ignored. This can lead to errors if the part of the sentence that is ignored is actually relevant, but in most cases, the approach works. This leads to a useful processing strategy in many applications, where the system only looks for information that is relevant to its task. This is in contrast to the syntactic parsing process that we reviewed previously, which required the system to analyze the entire sentence.

We can use the rule-based matcher in spaCy to create an application that analyzes a user's utterance to find values for these slots. The basic approach is to define patterns for the system to find words that specify a slot and to define corresponding tags that label the values with their slot names. The following code shows how to find some of the slots shown in *Figure 8.3* in sentences (we won't show the code for all of the slots in order to keep the example short):

```
import spacy
from spacy.lang.en import English

nlp = English()

ruler = nlp.add_pipe("entity_ruler")
cuisine_patterns = [
    {"label": "CUISINE", "pattern": "italian"},
    {"label": "CUISINE", "pattern": "german"},
    {"label": "CUISINE", "pattern": "chinese"}]
price_range_patterns = [
    {"label": "PRICE_RANGE", "pattern": "inexpensive"},
    {"label": "PRICE_RANGE", "pattern": "reasonably priced"},
    {"label": "PRICE_RANGE", "pattern": "good value"}]
atmosphere_patterns = [
    {"label": "ATMOSPHERE", "pattern": "casual"},
    {"label": "ATMOSPHERE", "pattern": "nice"},
    {"label": "ATMOSPHERE", "pattern": "cozy"}]
location_patterns = [
    {"label": "LOCATION", "pattern": "near here"},
    {"label": "LOCATION", "pattern": "walking distance"},
    {"label": "LOCATION", "pattern": "close by"},
    {"label": "LOCATION", "pattern": "a short drive"}]

ruler.add_patterns(cuisine_patterns)
ruler.add_patterns(price_range_patterns)
ruler.add_patterns(atmosphere_patterns)
ruler.add_patterns(location_patterns)

doc = nlp("can you recommend a casual italian restaurant within
walking distance")
```

```
print([(ent.text, ent.label_) for ent in doc.ents])
[('casual', 'ATMOSPHERE'), ('italian', 'CUISINE'), ('walking
distance', 'LOCATION')]
```

The preceding code starts by importing spaCy and the information that we need for processing text in the English language. The rule processor is called `ruler` and is added as a stage in the NLP pipeline. We then define three cuisines (a real application would likely have many more) and label them `CUISINE`. Similarly, we define patterns for recognizing price ranges, atmospheres, and locations. These rules state that if the user's sentence contains a specific word or phrase, such as `near here`, the `LOCATION` slot should be filled with that word or phrase.

The next step is to add the patterns to the rule processor (`ruler`) and then run the NLP processor on a sample sentence, *can you recommend a casual italian restaurant within walking distance?*. This process applies the rules to the document, which results in a set of labeled slots (which are called **entities** in spaCy) in `doc.ents`. By printing the slots and values, we can see that the processor found three slots, `ATMOSPHERE`, `CUISINE`, and `LOCATION`, with values of `casual`, `Italian`, and `walking distance`, respectively. By trying other sentences, we can confirm the following important characteristics of this approach to slot filling:

- Parts of the sentence that don't match a pattern, such as *can you recommend*, are ignored. This also means that the non-matching parts of the sentence can be nonsense, or they could actually be important to the meaning, but when they're ignored, the system can potentially make a mistake. For example, if the utterance were *can you recommend a casual non-Italian restaurant within walking distance*, the system would incorrectly think that the user wanted to find an Italian restaurant by using these rules. Additional rules can be written to take these kinds of cases into account, but in many applications, we just want to accept some inaccuracies as the price of keeping the application simple. This has to be considered on an application-by-application basis.

- Slots and values will be recognized wherever they occur in the sentences; they don't have to occur in any particular order.

- If a particular slot doesn't occur in the sentence, that doesn't cause any problems. It will just be left out of the resulting list of entities.

- If a slot occurs more than once, all occurrences will be recognized.

The names of the slot labels are up to the developer; they don't have to be specific values. For example, we could have said `TYPE_OF_FOOD` instead of `CUISINE`, and the processing would be the same.

We can use the spaCy visualizer, `displacy`, to get a clearer visualization of the result using the following code:

```
from spacy import displacy
colors = {"CUISINE": "#ea7e7e",
          "PRICE_RANGE": "#baffc9",
          "ATMOSPHERE": "#abcdef",
```

```
            "LOCATION": "#ffffba"}
options = {"ents": ["CUISINE","PRICE_RANGE","ATMOSPHERE","LOCATION"],
"colors": colors}
displacy.render(doc, style="ent", options=options, jupyter = True)
```

We can see the result in *Figure 8.4*, where the text and its slots and values are highlighted:

can you recommend a   casual **ATMOSPHERE**   italian **CUISINE**   within   walking distance **LOCATION**

Figure 8.4 – Slot visualization with displacy

Because our slots are custom (that is, not built into spaCy), to show colored slots, we have to define colors for the different slots (or `ents`) and assign the colors to the slots. We can then visualize the slots and values using a different color for each slot. The colors are defined in the `colors` variable in the preceding code. We can assign any colors to the slots that we find useful. The colors don't need to be different, but it is normally most helpful if they are, and if they are fairly distinctive. The values of the colors in this example are hexadecimal codes, which have standard color interpretations. `https://www.color-hex.com/` is a useful website that shows the hexadecimal values for many colors.

Using the spaCy id attribute

You have probably noticed that some of the slot values we have defined in this example mean the same thing – for example, `close by` and `near here`. If the slots and values are passed on to a later stage in processing such as database lookup, that later stage will have to have code for handling both `close by` and `near here`, even though the database lookup will be the same. This will complicate the application, so we would like to avoid it. spaCy provides another attribute of ent, `ent_id_`, for this purpose. This `id` attribute can be assigned in the patterns that find the slots, along with the label and pattern. This is accomplished by specifying an `id` attribute in the pattern declarations, which is a modification of the location patterns in the following code:

```
location_patterns = [
    {"label": "LOCATION", "pattern": "near here", "id":"nearby"},
    {"label": "LOCATION", "pattern": "close by","id":"nearby"},
    {"label": "LOCATION", "pattern": "near me","id":"nearby"},
    {"label": "LOCATION", "pattern": "walking distance", "id":"short_
walk"},
    {"label": "LOCATION", "pattern": "short walk", "id":"short_walk"},
    {"label": "LOCATION", "pattern": "a short drive", "id":"short_
drive"}]
```

If we print the slots, values, and IDs that result from *can you recommend a casual italian restaurant close by*, the result is as follows:

```
[('casual', 'ATMOSPHERE', ''), ('italian', 'CUISINE', ''), ('close
by', 'LOCATION', 'nearby')]
```

Here, we can see that the ID of `close by` is `nearby`, based on the `close by` pattern.

In the preceding code, we can see that the first three location patterns, which have similar meanings, have all been assigned the ID `nearby`. With this ID, the next stage in processing only needs to receive the `ent_id_` value, so it only has to handle `nearby`, and it doesn't have to have additional cases for `close by` and `near me`.

Note that in this example, the results for the `CUISINE` and `ATMOSPHERE` slots have empty IDs, because these were not defined in the `CUISINE` and `ATMOSPHERE` patterns. It is nevertheless good practice to define IDs for all patterns, if there are any IDs, in order to keep the results uniform.

Also note that these patterns reflect some design decisions about what phrases are synonymous and, therefore, should have the same ID, and which phrases are not synonymous and should have different IDs.

In the preceding code, we can see that `short walk` doesn't have the same ID as `near me`, for example. The design decision that was made here was to consider `short walk` and `near me` to have different meanings, and therefore to require different handling in the later stages of the application. Making decisions about which values are and are not synonymous will depend on the application and how rich the information available in the backend application is.

We have described several useful rule-based approaches to NLP. *Table 8.3* summarizes these rule-based techniques by listing three important properties of rule-based techniques:

- The format of the rules

- The types of processing that apply the rules to text

- How the results are represented

Name	Purpose	Model format	Processing	Result
Regular expressions	Recognize and format fixed phrases such as dates, times, addresses, and phone numbers	Regular expression syntax	Regular expression parser	Match object, string
Syntactic parsing	Recognize and label grammatical structures	Context-free grammar	Parsing (for example, chart parsing)	Constituent tree, dependency graph
Semantic analysis	Identify semantic relationships among elements of a sentence	Pattern rules	Semantic parsing	Slot-value pairs (slots are also referred to as "entities")

Table 8.3 – Formats, processing, and results for rule-based techniques

Summary

In this chapter, we've learned several important skills that make use of rules for processing natural language.

We've learned how to apply regular expressions to identify fixed-format expressions such as numbers, dates, and addresses. We've also learned about the uses of rule-based Python tools such as the NLTK syntactic parsing libraries for analyzing the syntactic structure of sentences and how to apply them. Finally, we've learned about rule-based tools for semantics analysis such as spaCy's `entity_ruler` for analyzing the slot-value semantics of sentences.

The next chapter, *Chapter 9*, will begin the discussion on machine learning by introducing statistical techniques such as classification with Naïve Bayes and **term frequency-inverse document frequency (TF-IDF)**, **support vector machines (SVMs)**, and conditional random fields. In contrast to the rule-based approaches we have discussed in this chapter, statistical approaches are based on models that are learned from training data and then applied to new, previously unseen data. Unlike the all-or-none rule-based systems, statistical systems are based on probabilities.

As we explore these techniques, we'll also consider how they can be combined with the rule-based techniques discussed in this chapter to create even more powerful and effective systems.

9

Machine Learning Part 1 – Statistical Machine Learning

In this chapter, we will discuss how to apply classical statistical machine learning techniques such as **Naïve Bayes**, **term frequency-inverse document frequency** (TF-IDF), **support vector machines** (**SVMs**), and **conditional random fields** (**CRFs**) to common **natural language processing** (**NLP**) tasks such as classification (or intent recognition) and slot filling.

There are two aspects of these classical techniques that we need to consider: representations and models. **Representation** refers to the format of the data that we are going to analyze. You will recall from *Chapter 7*, that it is standard to represent NLP data in formats other than lists of words. Numeric data representation formats such as vectors make it possible to use widely available numeric processing techniques, and consequently open up many possibilities for processing. In *Chapter 7*, we also explored data representations such as the **count bag of words**(**BoW**), TF-IDF, and Word2Vec. We will primarily be using TF-IDF in this chapter.

Once the data is in a format that is ready for further processing, that is, once it has been *vectorized*, we can use it to train, or build, a model that can be used to analyze similar data that the system may encounter in the future. This is the **training** phase. The future data can be test data; that is, previously unseen data similar to the training data that is used to evaluate the model. In addition, if the model is being used in a practical application, the future data could be new examples of queries from users or customers addressed to the runtime system. When the system is used to process data after the training phase, this is called **inference**.

We will cover the following topics in this chapter:

- A quick overview of evaluation
- Representing documents with TF-IDF and classifying with naïve Bayes
- Classifying documents with SVMs
- Slot filling with conditional random fields

We will start this chapter with a very practical and basic set of techniques that should be in everyone's toolbox, and that frequently end up being practical solutions to classification problems.

A quick overview of evaluation

Before we look at how different statistical techniques work, we have to have a way to measure their performance, and there are a couple of important considerations that we should review first. The first consideration is the *metric* or score that we assign to the system's processing. The most common and simple metric is **accuracy**, which is the number of correct responses divided by the overall number of attempts. For example, if we're attempting to measure the performance of a movie review classifier, and we attempt to classify 100 reviews as positive or negative, if the system classifies 75 reviews correctly, the accuracy is 75%. A closely related metric is **error rate**, which is, in a sense, the opposite of accuracy because it measures how often the system made a mistake. In this example, the error rate would be 25%.

We will only make use of accuracy in this chapter, although there are more precise and informative metrics that are actually more commonly used, for example, **precision**, **recall**, **F1**, and **area under the curve** (AUC). We will discuss these in *Chapter 13*. For the purposes of this chapter, we just need a basic metric so that we can compare results, and accuracy will be adequate for that.

The second important consideration that we need to keep in mind in evaluation is how to treat the data that we will be using for evaluation. Machine learning approaches are trained and evaluated with a standard approach that involves splitting the dataset into *training*, *development*, also often called *validation* data, and *testing* subsets. The training set is the data that is used to build the model, and is typically about 60-80 percent of the available data, although the exact percentage is not critical. Typically, you would want to use as much data as possible for training, while reserving a reasonable amount of data for evaluation purposes. Once the model is built, it can be tested with development data, normally about 10-20 percent of the overall dataset. Problems with the training algorithm are generally uncovered by trying to use the model on the development set. A final evaluation is done once, on the remaining data – the test data, which is usually about 10 percent of the total data.

Again, the exact breakdown of training, development, and test data is not critical. Your goal is to use the training data to build a good model that will enable your system to accurately predict the interpretation of new, previously unseen data. To meet that goal, you need as much training data as possible. The goal of the test data is to get an accurate measure of how your model performs on new data. To meet that goal, you need as much test data as possible. So, the data split will always involve a trade-off between these goals.

It is very important to keep the training data separate from the development data, and especially from the test data. Performance on the training data is not a good indicator of how the system will really perform on new, unseen data, so performance on the training data is not used for evaluation.

Following this brief introduction to evaluation, We will now move on to the main topics of this chapter. We will cover some of the most well-established machine learning approaches for important NLP applications such as classification and slot filling.

Representing documents with TF-IDF and classifying with Naïve Bayes

In addition to evaluation, two important topics in the general paradigm of machine learning are representation and processing algorithms. Representation involves converting a text, such as a document, into a numerical format that preserves relevant information about the text. This information is then analyzed by the processing algorithm to perform the NLP application. You've already seen a common approach to representation, TF-IDF, in *Chapter 7*. In this section, we will cover using TF-IDF with a common classification approach, Naïve Bayes. We will explain both techniques and show an example.

Summary of TF-IDF

You will recall the discussion of TF-IDF from *Chapter 7*. TF-IDF is based on the intuitive goal of trying to find words in documents that are particularly diagnostic of their classification topic. Words that are relatively infrequent in the whole corpus, but which are relatively common in a specific document, seem to be helpful in finding the document's class. TF-IDF was defined in the equations presented in the *term frequency-inverse document frequency (TF-IDF)* section in *Chapter 7*. In addition, we saw partial TF-IDF vectors for some of the documents in the movie review corpus in *Figure 7.4*. Here, we will take the TF-IDF vectors and use them to classify documents using the Naïve Bayes classification approach.

Classifying texts with Naïve Bayes

Bayesian classification techniques have been used for many years, and, despite their long history, are still very common and widely used. Bayesian classification is simple and fast, and can lead to acceptable results in many applications.

The formula for Bayesian classification is shown in the following equation. For each category in the set of possible categories, and for each document, we want to compute the probability that that is the correct category for that document. This computation is based on some representation of the document; in our case, the representation will be a vector like one of the ones we've previously discussed – BoW, TF-IDF, or Word2Vec.

To compute this probability, we take into account the probability of the vector, given a category, multiplied by the probability of the category, and divided by the probability of the document vector, as shown in the following equation:

$$P(category \mid documentVector) = \frac{P(documentVector \mid category)P(category)}{P(documentVector)}$$

The training process will determine the probabilities of the document vectors in each category and the overall probabilities of the categories.

The formula is called *naïve* because it makes the assumption that the features in the vectors are independent. This is clearly not correct for text because the words in sentences are not at all independent. However, this assumption makes the processing much simpler, and in practice does not usually make a significant difference in the results.

There are both binary and multi-class versions of Bayesian classification. We will work with binary classification with the movie review corpus since we have only two categories of reviews.

TF-IDF/Bayes classification example

Using TF-IDF and Naïve Bayes to classify the movie reviews, we can start by reading the reviews and splitting the data into training and test sets, as shown in the following code:

```
import sklearn
import os
from sklearn.feature_extraction.text import TfidfVectorizer
import nltk
from sklearn.datasets import load_files
path = './movie_reviews/'
# we will consider only the most 1000 common words
max_tokens = 1000


# load files -- there are 2000 files
movie_reviews = load_files(path)
# the names of the categories (the labels) are automatically generated
from the names of the folders in path
# 'pos' and 'neg'
labels = movie_reviews.target_names


# Split data into training and test sets
# since this is just an example, we will omit the dev test set
# 'movie_reviews.data' is the movie reviews
# 'movie_reviews.target' is the categories assigned to each review
# 'test_size = .20' is the proportion of the data that should be
reserved for testing
# 'random_state = 42' is an integer that controls the randomization of
the
# data so that the results are reproducible
from sklearn.model_selection import train_test_split
movies_train, movies_test, sentiment_train, sentiment_test = train_
test
_split(movie_reviews.data,
                        movie_reviews.target,
```

```
                                                             test_size
    = 0.20,

                    random_state = 42)
```

Once we have our training and test data, the next step is creating TF-IDF vectors from the reviews, as shown in the following code snippet. We will primarily be using the scikit-learn library, although we'll use NLTK for tokenization:

```
# initialize TfidfVectorizer to create the tfIdf representation of the
corpus
# the parameters are: min_df -- the percentage of documents that the
word has
# to occur in to be considered, the tokenizer to use, and the maximum
# number of words to consider (max_features)

vectorizer = TfidfVectorizer(min_df = .1,
                             tokenizer = nltk.word_tokenize,
                             max_features = max_tokens)

# fit and transform the text into tfidf format, using training text
# here is where we build the tfidf representation of the training data
movies_train_tfidf = vectorizer.fit_transform(movies_train)
```

The main steps in the preceding code are creating the vectorizer and then using the vectorizer to convert the movie reviews to TF-IDF format. This is the same process that we followed in *Chapter 7*. The resulting TF-IDF vectors were shown in *Figure 7.4*, so we won't repeat that here.

Classifying the documents into positive and negative reviews is then done with the multinomial Naïve Bayes function from scikit-learn, which is one of scikit-learn's Naïve Bayes packages, suitable for working with TF-IDF vector data. You can take a look at scikit-learn's other Naïve Bayes packages at https://scikit-learn.org/stable/modules/naive_bayes.html#naive-bayes for more information.

Now that we have the TF-IDF vectors, we can initialize the naïve Bayes classifier and train it on the training data, as shown here:

```
from sklearn.naive_bayes import MultinomialNB
# Initialize the classifier and train it
classifier = MultinomialNB()
classifier.fit(movies_train_tfidf, sentiment_train)
```

Finally, we can compute the accuracy of the classifier by vectorizing the test set (`movies_test_tfidf`), and using the classifier that was created from the training data to predict the classes of the test data, as shown in the following code:

```
# find accuracy based on test set
movies_test_tfidf = vectorizer.fit_transform(movies_test)
# for each document in the test data, use the classifier to predict
whether its sentiment is positive or negative
sentiment_pred = classifier.predict(movies_test_tfidf)
sklearn.metrics.accuracy_score(sentiment_test,
    sentiment_pred)
0.64
# View the results as a confusion matrix
from sklearn.metrics import confusion_matrix
conf_matrix = confusion_matrix(sentiment_test,
    sentiment_pred,normalize=None)
print(conf_matrix)
 [[132  58]
  [ 86 124]]
```

We can see from the preceding code that the accuracy of the classifier is `0.64`. That is, 64% of the test data reviews were assigned the correct category (positive or negative). We can also get some more information about how well the classification worked by looking at the *confusion matrix*, which is shown in the last two lines of the code.

In general, a confusion matrix shows which categories were confused with which other categories. We have a total of `400` test items (20% of the 2,000 reviews that were reserved as test examples). `132` of the `190` negative reviews were correctly classified as negative, and `58` were incorrectly classified as positive. Similarly, `124` of the `210` positive reviews were correctly classified as positive, but `86` were misclassified as negative. That means 69% of the negative reviews were correctly classified, and 59% of the positive reviews were correctly classified. From this, we can see that our model does slightly better in classifying negative reviews as negative. The reasons for this difference are not clear. To understand this result better, we can look more carefully at the reviews that were misclassified. We won't do this right now, but we will look at analyzing results more carefully in *Chapter 14*.

We will now consider a more modern and generally more accurate approach to classification.

Classifying documents with Support Vector Machines (SVMs)

SVMs are a popular and robust tool for text classification in applications such as intent recognition and chatbots. Unlike neural networks, which we will discuss in the next chapter, the training process is usually relatively quick and normally doesn't require enormous amounts of data. That means that SVMs are good for applications that have to be quickly deployed, perhaps as a preliminary step in the development of a larger-scale application.

The basic idea behind SVMs is that if we represent documents as **n**-dimensional vectors (for example, the TF-IDF vectors that we discussed in *Chapter 7*, we want to be able to identify a hyperplane that provides a boundary that separates the documents into two categories with as large a boundary (or *margin*) as possible.

An illustration of using SVMs on the movie review data is shown here. We start, as usual, by importing the data and creating a train/test split:

```
import numpy as np
from sklearn.datasets import load_files
from sklearn.svm import SVC
from sklearn.pipeline import Pipeline
# the directory root will be wherever the movie review data is located
directory_root = "./lab/movie_reviews/"
movie_reviews = load_files(directory_root,
    encoding='utf-8',decode_error="replace")
# count the number of reviews in each category
labels, counts = np.unique(movie_reviews.target,
    return_counts=True)
# convert review_data.target_names to np array
labels_str = np.array(movie_reviews.target_names)[labels]
print(dict(zip(labels_str, counts)))
{'neg': 1000, 'pos': 1000}
from sklearn.model_selection import train_test_split
movies_train, movies_test, sentiment_train, sentiment_test
    = train_test_split(movie_reviews.data,
        movie_reviews.target, test_size = 0.20,
        random_state = 42)
```

Now that we have the data set up and we have generated the train/test split, we will create the TF-IDF vectors and perform the SVC classification in the following code:

```
# We will work with a TF_IDF representation, as before
from sklearn.feature_extraction.text import TfidfVectorizer
```

```
from sklearn.metrics import classification_report, accuracy_score
# Use the Pipeline function to construct a list of two processes
# to run, one after the other -- the vectorizer and the classifier
svc_tfidf = Pipeline([
        ("tfidf_vectorizer", TfidfVectorizer(
        stop_words = "english", max_features=1000)),
        ("linear svc", SVC(kernel="linear"))
    ])
model = svc_tfidf
model.fit(movies_train, sentiment_train)
sentiment_pred = model.predict(movies_test)
accuracy_result = accuracy_score( sentiment_test,
    sentiment_pred)
print(accuracy_result)
0.8125
# View the results as a confusion matrix
from sklearn.metrics import confusion_matrix
conf_matrix = confusion_matrix(sentiment_test,
    sentiment_pred,normalize=None)
print(conf_matrix)
[[153  37]
 [ 38 172]]
```

The process shown here is very similar to the Naïve Bayes classification example shown in the code snippets in the previous section. However, in this case, we use an SVM instead of Naïve Bayes for classification, although we still use TF-IDF to vectorize the data. In the preceding code, we can see that the accuracy result for the classification is 0.82, which is considerably better than the Bayes accuracy shown in the previous section.

The resulting confusion matrix is also better, in that 153 of the 190 negative reviews were correctly classified as negative, and 37 were incorrectly classified as positive. Similarly, 172 of the 210 positive reviews were correctly classified as positive, but 38 were misclassified as negative. That means 80% of the negative reviews were correctly classified, and 81% of the positive reviews were correctly classified.

SVMs were originally designed for binary classification, as we have just seen with the movie review data, where we only have two categories (positive and negative, in this case). However, they can be extended to multi-class problems (which includes, most cases of intent recognition) by splitting the problem into multiple binary problems.

Consider a multi-class problem that might occur with a generic personal assistant application. Suppose the application includes several intents, for example:

- Find out the weather
- Play music

- Read the latest headlines

- Tell me the latest sports scores for my favorite teams

- Find a nearby restaurant offering a particular cuisine

The application needs to classify the user's query into one of these intents in order to process it and answer the user's question. To use SVMs for this, it is necessary to recast the problem as a set of binary problems. There are two ways to do this.

One is to create multiple models, one for each pair of classes, and split the data so that each class is compared to every other class. With the personal assistant example, the classification would need to decide about questions such as "*Is the category weather or sports?*" and *Is the category weather or news?*. This is the *one versus one* approach, and you can see that this could result in a very large number of classifications if the number of intents is large.

The other approach is called *one versus rest* or *one versus all*. Here, the idea is to ask questions such as *Is the category "weather" or is it something else?* This is the more popular approach, which we will show here.

The way to use the multiclass SVM from scikit-learn is very similar to what was shown previously. The difference is that we're importing the `OneVsRestClassifier` and using it to create the classification model, as shown in the following code:

```
from sklearn.multiclass import OneVsRestClassifier
model = OneVsRestClassifier(SVC())
```

Classification is widely used in many natural language applications, both for categorizing documents such as movie reviews and for categorizing what a user's overall goal (or *intent*) in asking their question is in applications such as chatbots. However, often the application will require more fine-grained information from the utterance or document in addition to its overall classification. This process is often called **slot filling**. We discussed slot-filling in *Chapter 8* and showed how to write slot-filling rules. In the following section, we will show another approach to slot-filling, based on statistical techniques, specifically **conditional random fields (CRFs)**.

Slot-filling with CRFs

In *Chapter 8*, we discussed the popular application of slot-filling, and we used the spaCy rule engine to find slots for the restaurant search application shown in *Figure 8.9*. This required writing rules for finding the fillers of each slot in the application. This approach can work fairly well if the potential slot fillers are known in advance, but if they aren't known in advance, it won't be possible to write rules. For example, with the rules in the code following *Figure 8.9*, if a user asked for a new cuisine, say, *Thai*, the rules wouldn't be able to recognize *Thai* as a new filler for the CUISINE slot, and wouldn't be able to recognize *not too far away* as a filler for the LOCATION slot. Statistical methods, which we will discuss in this section, can help with this problem.

With **statistical methods**, the system does not use rules but looks for patterns in its training data that can be applied to new examples. Statistical methods depend on having enough training data for the system to be able to learn accurate patterns, but if there is enough training data, statistical methods will generally provide a more robust solution to an NLP problem than rule-based methods.

In this section, we will look at a popular approach that can be applied to statistical slot filling – CRF. CRFs are a way of taking the context of textual items into account when trying to find the label for a span of text. Recall that the rules we discussed in *Chapter 8* did not look at any nearby words or other context – they only looked at the item itself. In contrast, CRFs attempt to model the probability of a label for a particular section of text, that is, given an input x, they model the probability of that input being an example category y $(P(y|x))$. CRFs use the word (or token) sequence to estimate the conditional probabilities of slot labels in that context. We will not review the mathematics of CRF here, but you can find many detailed descriptions of the underlying mathematics on the web, for example, `https://arxiv.org/abs/1011.4088`.

To train a system for slot-tagging, the data has to be annotated so that the system can tell what slots it's looking for. There are at least four different formats used in the NLP technical literature for representing the annotations for slot-tagged data, and we'll briefly illustrate these. These formats can be used both for training data and for representing processed NLP results, which can, in turn, be used for further processing stages such as retrieving information from a database.

Representing slot-tagged data

Training data for slot-filling applications can be found in several formats. Let's look at how the sentence `show me science fiction films directed by steven spielberg`, a query from the MIT movie query corpus (`https://groups.csail.mit.edu/sls/downloads/`), can be represented in four different formats.

One commonly used notation is **JavaScript Object Notation (JSON)** format, as shown here:

```
{tokens": "show me science fiction films directed by steven spielberg"
"entities": [
  {"entity": {
    "tokens": "science fiction films",
    "name": "GENRE"
  }},
  {
  "entity": {
   "tokens": "steven spielberg",
   "name": "DIRECTOR"
  }}
  ]
  }
```

Here, we see that the input sentence is shown as `tokens`, which is followed by a list of slots (called `entities` here). Each entity is associated with a name, such as `GENRE` or `DIRECTOR` and the tokens that it applies to. The example shows two slots, `GENRE` and `DIRECTOR`, which are filled by `science fiction films` and `steven spielberg`, respectively:

The second format uses **Extensible Markup Language (XML)** tags to label the slots, as in `show me <GENRE>science fiction films</GENRE> directed by <DIRECTOR>steven Spielberg</DIRECTOR>`.

The third format is called **Beginning Inside Outside (BIO)**, which is a textual format that labels the beginning, middle, and end of each slot filler in a sentence, as in *Figure 9.1*:

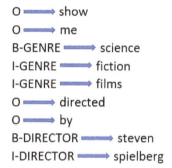

Figure 9.1 – BIO tagging for "show me science fiction films directed by steven spielberg"

In the BIO format, words outside of any slot are labeled O (that is `Outside`), the beginning of a slot (`science` and `steven` are labeled B, and the inside of a slot is labeled I).

Finally, another very simple format for representing tagged slots is **Markdown**, a simplified way of marking up text for rendering. We've seen Markdown before, in Jupyter notebooks, as a way to display comment blocks. In *Figure 9.2*, we can see an example of some training data for a restaurant search application similar to the one we looked at in *Chapter 8* (which was shown in *Figure 8.9*). Slot values are shown in square brackets, and the slot names are shown in parentheses:

```
## restaurant_search

- I want to get some [lunch](meal)

- I am searching for a [dinner](meal) spot

- I'm looking for a place in the [north](location) of town

- show me some [good](quality) [Chinese](cuisine) restaurants in the [north](location)

- how about a [Mexican](cuisine) restaurant [downtown](location)

- Are there any [Indian](cuisine) spots near here

- Italian restaurants on the [west side] (location)

- looking for [German](cuisine) places in the [south](location)

- what [Greek](cuisine) places are near [12345](location)

- help me find a [casual](atmosphere)[Asian fusion](cuisine) place

- I am looking for a [French](cuisine) restaurant [nearby](location)

- I am looking for a [nice] (quality) [Mexican](cuisine) or [thai](cuisine) place that's [not too
expensive](price)

- [cozy](atmosphere) [barbecue](cuisine) restaurant
```

Figure 9.2 – Markdown tagging for a restaurant search application

All four formats show basically the same information about the slots and values; the information is just represented in slightly different ways. For your own projects, if you are using a public dataset, it would probably be most convenient to use the format that the dataset already uses. However, if you are using your own data, you can choose whatever format is most appropriate or easiest to use for your application. All other considerations being equal, the XML and JSON formats are generally more flexible than BIO or markdown, since they can represent nested slots, that is, slots that contain additional values as slot fillers.

For our example, we will be using the spaCy CRF suite library, located at `https://github.com/ talmago/spacy_crfsuite`, and we will use restaurant search as an example application. This dataset is annotated using the Markdown format.

The following code sets up the application by importing display and Markdown functionality and then reading in the Markdown file from the `examples` directory. Reading in the Markdown file will reproduce the list of utterances shown in *Figure 9.2*. Note that the training data in the Markdown file is not large enough for a real application, but it works as an example here:

```
from IPython.display import display, Markdown
with open("examples/restaurant_search.md", "r") as f:
    display(Markdown(f.read()))
```

The next steps, shown below, will be to import `crfsuite` and `spacy` and convert the Markdown-formatted training dataset to a CRF format. (The code in GitHub shows some additional steps that are omitted here for simplicity):

```
import sklearn_crfsuite
from spacy_crfsuite import read_file

train_data = read_file("examples/restaurant_search.md")
train_data
In [ ]:
import spacy
from spacy_crfsuite.tokenizer import SpacyTokenizer
from spacy_crfsuite.train import gold_example_to_crf_tokens
nlp = spacy.load("en_core_web_sm", disable=["ner"])
tokenizer = SpacyTokenizer(nlp)
train_dataset = [
    gold_example_to_crf_tokens(ex, tokenizer=tokenizer)
    for ex in train_data
]
train_dataset[0]
```

At this point, we can do the actual training of the CRF with the `CRFExtractor` object, as shown here:

```
from spacy_crfsuite import CRFExtractor

crf_extractor = CRFExtractor(
    component_config=component_config)
crf_extractor

rs = crf_extractor.fine_tune(train_dataset, cv=5,
    n_iter=50, random_state=42)
print("best_params:", rs.best_params_, ", score:",
    rs.best_score_)
crf_extractor.train(train_dataset)

classification_report = crf_extractor.eval(train_dataset)
print(classification_report[1])
```

The classification report, which is produced in the second to last step, and is shown next, is based on the training dataset (`train_dataset`). Since the CRF was trained on this dataset, the classification report will show perfect performance on each slot. Obviously, this is not realistic, but it's shown here in order to illustrate the classification report. Remember that we will return to the topics of precision, recall, and F1 score in *Chapter 13*:

```
                precision    recall  f1-score    support
```

U-atmosphere	1.000	1.000	1.000	1
U-cuisine	1.000	1.000	1.000	9
U-location	1.000	1.000	1.000	6
U-meal	1.000	1.000	1.000	2
B-price	1.000	1.000	1.000	1
I-price	1.000	1.000	1.000	1
L-price	1.000	1.000	1.000	1
U-quality	1.000	1.000	1.000	1
micro avg	1.000	1.000	1.000	22
macro avg	1.000	1.000	1.000	22
weighted avg	1.000	1.000	1.000	22

At this point, the CRF model has been trained and is ready to test with new data. If we test this model with the sentence *show me some good chinese restaurants near me*, we can see the result in JSON format in the following code. The CRF model found two slots, CUISINE and QUALITY, but missed the LOCATION slot, which should have been filled by near me. The results also show the model's confidence in the slots, which was quite high, well over 0.9. The result also includes the zero-based positions of the characters in the input that begin and end the slot value (good starts at position 10 and ends at position 14):

```
example = {"text": "show some good chinese restaurants near me"}
tokenizer.tokenize(example, attribute="text")
crf_extractor.process(example)
[{'start': 10,
  'end': 14,
  'value': 'good',
  'entity': 'quality',
  'confidence': 0.9468721304898786},
 {'start': 15,
  'end': 22,
  'value': 'chinese',
  'entity': 'cuisine',
  'confidence': 0.9591743424660175}]
```

Finally, we can illustrate the robustness of this approach by testing our application with a cuisine that was not seen in the training data, Japanese. Let's see if the system can label Japanese as a cuisine in a new utterance. We can try an utterance such as show some good Japanese restaurants near here and see the result in the following JSON:

```
[{'start': 10,
  'end': 14,
  'value': 'good',
```

```
 'entity': 'quality',
 'confidence': 0.6853277275481114},
{'start': 15,
 'end': 23,
 'value': 'japanese',
 'entity': 'cuisine',
 'confidence': 0.537198793062902}]
```

The system did identify Japanese as a cuisine in this example, but the confidence was much lower than we saw in the previous example, only 0.537 this time, compared with 0.96 for the same sentence with a known cuisine. This relatively low confidence is typical for slot fillers that didn't occur in the training data. Even the confidence of the QUALITY slot (which did occur in the training data) was lower, probably because it was affected by the low probability of the unknown CUISINE slot filler.

A final observation worth pointing out is that while it would have been possible to develop a rule-based slot tagger for this task, as we saw in *Chapter 8*, the resulting system would not have been able to even tentatively identify Japanese as a slot filler unless Japanese had been included in one of the rules. This is a general illustration of how the statistical approach can provide results that are not all or none, compared to rule-based approaches.

Summary

This chapter has explored some of the basic and most useful classical statistical techniques for NLP. They are especially valuable for small projects that start out without a large amount of training data, and for the exploratory work that often precedes a large-scale project.

We started out by learning about some basic evaluation concepts. We learned particularly about accuracy, but we also looked at some confusion matrices. We also learned how to apply Naïve Bayes classification to texts represented in TF-IDF format, and then we worked through the same classification task using a more modern technique, SVMs. Comparing the results produced by Naïve Bayes and SVMs, we saw that we got better performance from the SVMs. We then turned our attention to a related NLP task, slot-filling. We learned about different ways to represent slot-tagged data and finally illustrated CRFs with a restaurant recommendation task. These are all standard approaches that are good to have in your NLP toolbox, especially for the initial exploration of applications with limited data.

In *Chapter 10*, we will continue working on topics in machine learning, but we will move on to a very different type of machine learning, neural networks. There are many varieties of neural networks, but overall, neural networks and their variants have become the standard technologies for NLP in the last decade or so. The next chapter will introduce this important topic.

10

Machine Learning Part 2 – Neural Networks and Deep Learning Techniques

Neural networks (**NNs**) have only became popular in **natural language understanding** (**NLU**) around 2010 but have since been widely applied to many problems. In addition, there are many applications of NNs to non-**natural language processing** (**NLP**) problems such as image classification. The fact that NNs are a general approach that can be applied across different research areas has led to some interesting synergies across these fields.

In this chapter, we will cover the application of **machine learning** (**ML**) techniques based on NNs to problems such as NLP classification. We will also cover several different kinds of commonly used NNs—specifically, fully connected **multilayer perceptrons** (**MLPs**), **convolutional NNs** (**CNNs**), and **recurrent NNs** (**RNNs**)—and show how they can be applied to problems such as classification and information extraction. We will also discuss fundamental NN concepts such as hyperparameters, learning rate, activation functions, and epochs. We will illustrate NN concepts with a classification example using the TensorFlow/Keras libraries.

In this chapter, we will cover the following topics:

- Basics of NNs
- Example—MLP for classification
- Hyperparameters and tuning
- Moving beyond MLPs—RNNs
- Looking at another approach—CNNs

Basics of NNs

The basic concepts behind NNs have been studied for many years but have only fairly recently been applied to NLP problems on a large scale. Currently, NNs are one of the most popular tools for solving NLP tasks. NNs are a large field and are very actively researched, so we won't be able to give you a comprehensive understanding of NNs for NLP. However, we will attempt to provide you with some basic knowledge that will let you apply NNs to your own problems.

NNs are inspired by some properties of the animal nervous system. Specifically, animal nervous systems consist of a network of interconnected cells, called *neurons*, that transmit information throughout the network with the result that, given an input, the network produces an output that represents a decision about the input.

Artificial NNs (ANNs) are designed to model this process in some respects. The decision about how to react to the inputs is determined by a sequence of processing steps starting with units (*neurons*) that receive inputs and create outputs (or *fire*) if the correct conditions are met. When a neuron fires, it sends its output to other neurons. These next neurons receive inputs from a number of other neurons, and they in turn fire if they receive the right inputs. Part of the decision process about whether to fire involves *weights* on the neurons. The way that the NN learns to do its task—that is, the *training process*—is the process of adjusting the weights to produce the best results on the training data.

The training process consists of a set of *epochs*, or passes through the training data, adjusting the weights on each pass to try to reduce the discrepancy between the result produced by the NN and the correct results.

The neurons in an NN are arranged in a series of layers, with the final layer—the output layer—producing the decision. Applying these concepts to NLP, we will start with an input text that is fed to the input layer, which represents the input being processed. Processing proceeds through all the layers, continuing through the NN until it reaches the output layer, which provides the decision—for example, is this movie review positive or negative?

Figure 10.1 represents a schematic diagram of an NN with an input layer, two hidden layers, and an output layer. The NN in *Figure 10.1* is a **fully connected NN (FCNN)** because every neuron receives inputs from every neuron in the preceding layer and sends outputs to every neuron in the following layer:

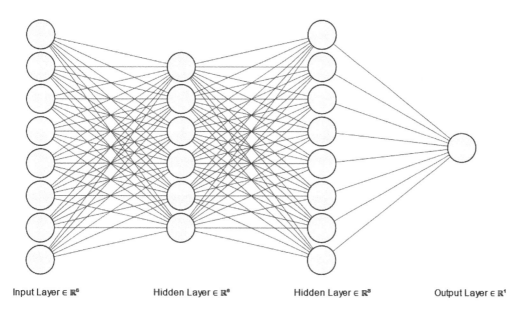

Input Layer ∈ ℝ⁶   Hidden Layer ∈ ℝ⁶   Hidden Layer ∈ ℝ⁹   Output Layer ∈ ℝ¹

Figure 10.1 – An FCNN with two hidden layers

The field of NNs uses a lot of specialized vocabulary, which can sometimes make it difficult to read documentation on the topic. In the following list, we'll provide a brief introduction to some of the most important concepts, referring to *Figure 10.1* as needed:

- **Activation function**: The activation function is the function that determines when a neuron has enough inputs to fire and transmit its output to the neurons in the next layer. Some common activation functions are sigmoid and **rectified linear unit (ReLU)**.

- **Backpropagation**: The process of training an NN where the loss is fed back through the network to train the weights.

- **Batch**: A batch is a set of samples that will be trained together.

- **Connection**: A link between neurons, associated with a weight that represents the strength of the connection. The lines between neurons in *Figure 10.1* are connections.

- **Convergence**: A network has converged when additional epochs do not appear to produce any reduction in the loss or improvements in accuracy.

- **Dropout**: A technique for preventing overfitting by randomly removing neurons.

- **Early stopping**: Ending training before the planned number of epochs because training appears to have converged.

- **Epoch**: One pass through the training data, adjusting the weights to minimize loss.

- **Error**: The difference between the predictions produced by an NN and the reference labels. Measures how well the network predicts the classification of the data.

- **Exploding gradients**: Exploding gradients occur when gradients become unmanageably large during training.

- **Forward propagation**: Propagation of inputs forward through an NN from the input layer through the hidden layers to the output layer.

- **Fully connected**: An FCNN is an NN "*with every neuron in one layer connecting to every neuron in the next layer*" (`https://en.wikipedia.org/wiki/Artificial_neural_network`), as shown in *Figure 10.1*.

- **Gradient descent**: Optimizing weights by adjusting them in a direction that will minimize loss.

- **Hidden layer**: A layer of neurons that is not the input or output layer.

- **Hyperparameters**: Parameters that are not learned and are usually adjusted in a manual tuning process in order for the network to produce optimal results.

- **Input layer**: The layer in an NN that receives the initial data. This is the layer on the left in *Figure 10.1*.

- **Layer**: A set of neurons in an NN that takes information from the previous layer and passes it on to the next layer. *Figure 10.1* includes four layers.

- **Learning**: Assigning weights to connections in the training process in order to minimize loss.

- **Learning rate/adaptive learning rate**: Amount of adjustment to the weights after each epoch. In some approaches, the learning rate can adapt as the training progresses; for example, if learning starts to slow, it can be useful to decrease the learning rate.

- **Loss**: A function that provides a metric that quantifies the distance between the current model's predictions and the goal values. The training process attempts to minimize loss.

- **MLP**: As described on *Wikipedia*, "*a fully connected class of feedforward artificial neural network (ANN). An MLP consists of at least three layers of nodes: an input layer, a hidden layer and an output layer*" (`https://en.wikipedia.org/wiki/Multilayer_perceptron`). *Figure 10.1* shows an example of an MLP.

- **Neuron (unit)**: A unit in an NN that receives inputs and computes outputs by applying an activation function.

- **Optimization**: Adjustment to the learning rate during training.

- **Output layer**: The final layer in an NN that produces a decision about the input. This is the layer on the right in *Figure 10.1*.

- **Overfitting**: Tuning the network too closely to the training data so that it does not generalize to previously unseen test or validation data.

- **Underfitting**: Underfitting occurs when an NN is unable to obtain good accuracy for training data. It can be addressed by using more training epochs or more layers.

- **Vanishing gradients**: Gradients that become so small that the network is unable to make progress.

- **Weights**: A property of the connection between neurons that represents the strength of the connection. Weights are learned during training.

In the next section, we will make these concepts concrete by going through an example of text classification with a basic MLP.

Example – MLP for classification

We will review basic NN concepts by looking at the MLP, which is conceptually one of the most straightforward types of NNs. The example we will use is the classification of movie reviews into reviews with positive and negative sentiments. Since there are only two possible categories, this is a *binary* classification problem. We will use the *Sentiment Labelled Sentences Data Set (From Group to Individual Labels using Deep Features, Kotzias et al., KDD 2015* https://archive.ics.uci.edu/ml/datasets/Sentiment+Labelled+Sentences), available from the University of California, Irvine. Start by downloading the data and unzipping it into a directory in the same directory as your Python script. You will see a directory called sentiment labeled sentences that contains the actual data in a file called imdb_labeled.txt. You can install the data into another directory of your choosing, but if you do, be sure to modify the filepath_dict variable accordingly.

You can take a look at the data using the following Python code:

```python
import pandas as pd
import os
filepath_dict = {'imdb':    'sentiment labelled sentences/imdb_
labelled.txt'}
document_list = []
for source, filepath in filepath_dict.items():
    document = pd.read_csv(filepath, names=['sentence', 'label'],
sep='\t')
    document['source'] = source
    document_list.append(document)

document = pd.concat(document_list)
print(document.iloc[0])
```

The output from the last print statement will include the first sentence in the corpus, its label (1 or 0—that is, positive or negative), and its source (Internet Movie Database IMDB).

In this example, we will vectorize the corpus using the scikit-learn count **bag of words** (**BoW**) vectorizer (CountVectorizer), which we saw earlier in *Chapter 7*.

The following code snippet shows the start of the vectorization process, where we set up some parameters for the vectorizer:

```
from sklearn.feature_extraction.text import CountVectorizer

# min_df is the minimum proportion of documents that contain the word
(excludes words that
# are rarer than this proportion)
# max_df is the maximum proportion of documents that contain the word
(excludes words that
# are rarer than this proportion
# max_features is the maximum number of words that will be considered
# the documents will be lowercased
vectorizer = CountVectorizer(min_df = 0, max_df = 1.0, max_features =
1000, lowercase = True)
```

The CountVectorizer function has some useful parameters that control the maximum number of words that will be used to build the model, as well as make it possible to exclude words that are considered to be too frequent or too rare to be very useful in distinguishing documents.

The next step is to do the train-test split, as shown in the following code block:

```
# split the data into training and test
from sklearn.model_selection import train_test_split

document_imdb = document[document['source'] == 'imdb']
reviews = document_imdb['sentence'].values
y = document_imdb['label'].values

# since this is just an example, we will omit the dev test set
# 'reviews.data' is the movie reviews
# 'y_train' is the categories assigned to each review in the training
data
# 'test_size = .20' is the proportion of the data that should be
reserved for testing
# 'random_state = 42' is an integer that controls the randomization of
the data so that the results are reproducible
reviews_train, reviews_test, y_train, y_test = train_test_split(
    reviews, y, test_size = 0.20, random_state = 42)
```

The preceding code shows the splitting of the training and test data, reserving 20% of the total data for testing.

The `reviews` variable holds the actual documents, and the `y` variable holds their labels. Note that X and y are frequently used in the literature to represent the data and the categories in an ML problem, respectively, although we're using `reviews` for the X data here:

```
vectorizer.fit(reviews_train)
vectorizer.fit(reviews_test)
X_train = vectorizer.transform(reviews_train)
X_test  = vectorizer.transform(reviews_test)
```

The preceding code shows the process of vectorizing the data, or converting each document to a numerical representation, using the vectorizer that was defined previously. You can review vectorization by going back to *Chapter 7*.

The result is `X_train`, the count BoW of the dataset. You will recall the count BoW from *Chapter 7*.

The next step is to set up the NN. We will be using the Keras package, which is built on top of Google's TensorFlow ML package. Here's the code we need to execute:

```
from keras.models import Sequential
from keras import layers
from keras import models

# Number of features (words)
# This is based on the data and the parameters that were provided to
the vectorizer
# min_df, max_df and max_features
input_dimension = X_train.shape[1]
print(input_dimension)
```

The code first prints the input dimension, which in this case is the number of words in each document vector. The input dimension is useful to know because it's computed from the corpus, as well as the parameters we set in the `CountVectorizer` function. If it is unexpectedly large or small, we might want to change the parameters to make the vocabulary larger or smaller.

The following code defines the model:

```
# a Sequential model is a stack of layers where each layer has one
input and one output tensor
# Since this is a binary classification problem, there will be one
output (0 or 1)
# depending on whether the review is positive or negative
# so the Sequential model is appropriate
model = Sequential()
model.add(layers.Dense(16, input_dim = input_dimension, activation =
'relu'))
model.add(layers.Dense(16, activation = 'relu'))
```

```
model.add(layers.Dense(16, activation = 'relu'))
# output layer
model.add(layers.Dense(1, activation = 'sigmoid'))
```

The model built in the preceding code includes the input layer, two hidden layers, and one output layer. Each call to the model.add() method adds a new layer to the model. All the layers are dense because, in this fully connected network, every neuron receives inputs from every neuron in the previous layer, as illustrated in *Figure 10.1*. The 2 hidden layers each contain 16 neurons. Why do we specify 16 neurons? There is no hard and fast rule for how many neurons to include in the hidden layers, but a general approach would be to start with a smaller number since the training time will increase as the number of neurons increases. The final output layer will only have one neuron because we only want one output for this problem, whether the review is positive or negative.

Another very important parameter is the **activation function**. The activation function is the function that determines how the neuron responds to its inputs. For all of the layers in our example, except the output layer, this is the ReLU activation function. The ReLU function can be seen in *Figure 10.2*. ReLU is a very commonly used activation function:

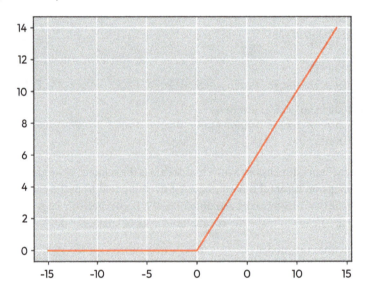

Figure 10.2 – Values of the ReLU function for inputs between -15 and 15

One of the most important benefits of the ReLU function is that it is very efficient. It has also turned out to generally give good results in practice and is normally a reasonable choice as an activation function.

The other activation function that's used in this NN is the sigmoid function, which is used in the output layer. We use the sigmoid function here because in this problem we want to predict the probability of a positive or negative sentiment, and the value of the sigmoid function will always be between 0 and 1. The formula for the sigmoid function is shown in the following equation:

$$S(x) \ = \ \frac{1}{1 + e^{-x}}$$

A plot of the sigmoid function is shown in *Figure 10.3*, and it is easy to see that its output value will always be between 0 and 1 regardless of the value of the input:

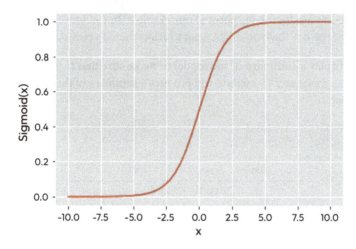

Figure 10.3 – Values of the sigmoid function for inputs between -10 and 10

The sigmoid and ReLU activation functions are popular and practical activation functions, but they are only two examples of the many possible NN activation functions. If you wish to investigate this topic further, the following *Wikipedia* article is a good place to start: https://en.wikipedia.org/wiki/Activation_function.

Once the model has been defined, we can compile it, as shown in the following code snippet:

```
model.compile(loss = 'binary_crossentropy',
              optimizer = 'adam',
              metrics = ['accuracy'])
```

The model.compile() method requires the loss, optimizer, and metrics parameters, which supply the following information:

- The loss parameter, in this case, tells the compiler to use binary_crossentropy to compute the loss. **Binary cross-entropy** is a commonly used loss function for binary problems such as binary classification. A similar function, categorical_crossentropy, is used for problems when there are two or more label classes in the output. For example, if the task were to assign a star rating to reviews, we might have five output classes corresponding to the five possible star ratings, and in that case, we would use categorical cross-entropy.

- The `optimizer` parameter adjusts the learning rate during training. We will not go into the mathematical details of `adam` here, but generally speaking, the optimizer we use here, `adam`, normally turns out to be a good choice.

- Finally, the `metrics` parameter tells the compiler how we will evaluate the quality of the model. We can include multiple metrics in this list, but we will just include `accuracy` for now. In practice, the metrics you use will depend on your problem and dataset, but `accuracy` is a good metric to use for the purposes of our example. In *Chapter 13*, we will explore other metrics and the reasons that you might want to select them in particular situations.

It is also helpful to display a summary of the model to make sure that the model is structured as intended. The `model.summary()` method will produce a summary of the model, as shown in the following code snippet:

```
model.summary()

Model: "sequential"
```

Layer (type)	Output Shape	Param #
dense (Dense)	(None, 16)	13952
dense_1 (Dense)	(None, 16)	272
dense_2 (Dense)	(None, 16)	272
dense_3 (Dense)	(None, 1)	17

```
Total params: 14,513
Trainable params: 14,513
Non-trainable params: 0
```

In this output, we can see that the network, consisting of four dense layers (which are the input layer, the two hidden layers, and the output layer), is structured as expected.

The final step is to fit or train the network, using the following code:

```
history = model.fit(X_train, y_train,
                    epochs=20,
                    verbose=True,
                    validation_data=(X_test, y_test),
                    batch_size=10)
```

Training is the iterative process of putting the training data through the network, measuring the loss, adjusting the weights to reduce the loss, and putting the training data through the network again. This step can be quite time-consuming, depending on the size of the dataset and the size of the model.

Each cycle through the training data is an epoch. The number of epochs in the training process is a *hyperparameter*, which means that it's adjusted by the developer based on the training results. For example, if the network's performance doesn't seem to be improving after a certain number of epochs, the number of epochs can be reduced since additional epochs are not improving the result. Unfortunately, there is no set number of epochs after which we can stop training. We have to observe the improvements in accuracy and loss over epochs to decide whether the system is sufficiently trained.

Setting the `verbose = True` parameter is optional but useful because this will produce a trace of the results after each epoch. If the training process is long, the trace can help you verify that the training is making progress. The batch size is another hyperparameter that defines how many data samples are to be processed before updating the model. When the following Python code is executed, with `verbose` set to `True`, at the end of every epoch, the loss, the accuracy, and the validation loss and accuracy will be computed. After training is complete, the `history` variable will contain information about the progress of the training process, and we can see plots of the training progress.

It is important to display how the plots of accuracy and loss change with each epoch because it will give us an idea of how many epochs are needed to get this training to converge and will make it very clear when the data is overfitting. The following code shows how to plot the accuracy and loss changes over epochs::

```
import matplotlib.pyplot as plt
plt.style.use('ggplot')
def plot_history(history):
    acc = history.history['accuracy']
    val_acc = history.history['val_accuracy']
    loss = history.history['loss']
    val_loss = history.history['val_loss']
    x = range(1, len(acc) + 1)

    plt.figure(figsize=(12, 5))
    plt.subplot(1, 2, 1)
    plt.plot(x, acc, 'b', label='Training accuracy')
    plt.plot(x, val_acc, 'r', label = 'Validation accuracy')
    plt.title('Training and validation accuracy')
    plt.legend()
    plt.subplot(1, 2, 2)
    plt.plot(x, loss, 'b', label='Training loss')
    plt.plot(x, val_loss, 'r', label='Validation loss')
    plt.title('Training and validation loss')
    plt.legend()
    plt.show()
plot_history(history)
```

We can see the results of the progress of our example through training over 20 epochs in *Figure 10.4*:

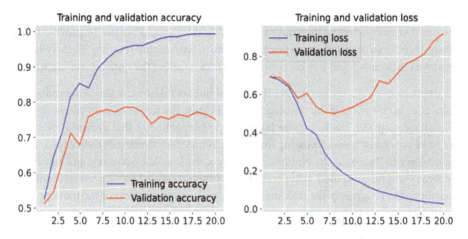

Figure 10.4 – Accuracy and loss over 20 epochs of training

Over the 20 epochs of training, we can see that the training accuracy approaches **1.0** and the training loss approaches **0**. However, this apparently good result is misleading because the really important results are based on the validation data. Because the validation data is not being used to train the network, it is the performance on the validation data that actually predicts how the network will perform in use. We can see from the plots of the changes in validation accuracy and loss that doing more training epochs after about 10 is not improving the model's performance on the validation data. In fact, it is increasing the loss and therefore making the model worse. This is clear from the increase in the validation loss in the graph on the right in *Figure 10.4*.

Improving performance on this task will involve modifying other factors, such as hyperparameters and other tuning processes, which we will go over in the next section.

Hyperparameters and tuning

Figure 10.4 clearly shows that increasing the number of training epochs is not going to improve performance on this task. The best validation accuracy seems to be about 80% after 10 epochs. However, 80% accuracy is not very good. How can we improve it? Here are some ideas. None of them is guaranteed to work, but it is worth experimenting with them:

- If more training data is available, the amount of training data can be increased.

- Preprocessing techniques that can remove noise from the training data can be investigated—for example, stopword removal, removing non-words such as numbers and HTML tags, stemming and lemmatization, and lowercasing. Details on these techniques were covered in *Chapter 5*.

- Changes to the learning rate—for example, lowering the learning rate might improve the ability of the network to avoid local minima.

- Decreasing the batch size.

- Changing the number of layers and the number of neurons in each layer is something that can be tried, but having too many layers is likely to lead to overfitting.

- Adding dropout by specifying a hyperparameter that defines the probability that the outputs from a layer will be ignored. This can help make the network more robust to overfitting.

- Improvements in vectorization—for example, by using **term frequency-inverse document frequency (TF-IDF)** instead of count BoW.

A final strategy for improving performance is to try some of the newer ideas in NNs—specifically, RNNs, CNNs, and transformers.

We will conclude this chapter by briefly reviewing RNNs and CNNs. We will cover transformers in *Chapter 11*.

Moving beyond MLPs – RNNs

RNNs are a type of NN that is able to take into account the order of items in an input. In the example of the MLP that was discussed previously, the vector representing the entire input (that is, the complete document) was fed to the NN at once, so the network had no way of taking into account the order of words in the document. However, this is clearly an oversimplification in the case of text data since the order of words can be very important to the meaning. RNNs are able to take into account the order of words by using earlier outputs as inputs to later layers. This can be especially helpful in certain NLP problems where the order of words is very important, such as **named entity recognition (NER)**, **part-of-speech (POS) tagging**, or **slot labeling**.

A diagram of a unit of an RNN is shown in *Figure 10.5*:

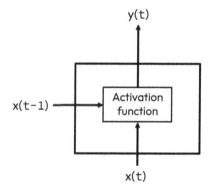

Figure 10.5 – A unit of an RNN

The unit is shown at time *t*. The input at time *t*, *x(t)*, is passed to the activation function as in the case of the MLP, but the activation function also receives the output from time *t-1*—that is, *x(t-1)*. For NLP, the earlier input would most likely have been the previous word. So, in this case, the input is the current word and one previous word. Using an RNN with Keras is very similar to the MLP example that we saw earlier, with the addition of a new RNN layer in the layer stack.

However, as the length of the input increases, the network will tend to *forget* information from earlier inputs, because the older information will have less and less influence over the current state. Various strategies have been designed to overcome this limitation, such as **gated recurrent units (GRUs)** and **long short-term memory (LSTM)**. If the input is a complete text document (as opposed to speech), we have access not only to previous inputs but also to future inputs, and a bidirectional RNN can be used.

We will not cover these additional variations of RNNs here, but they do often improve performance on some tasks, and it would be worth researching them. Although there is a tremendous amount of resources available on this popular topic, the following *Wikipedia* article is a good place to start: `https://en.wikipedia.org/wiki/Recurrent_neural_network`.

Looking at another approach – CNNs

CNNs are very popular for image recognition tasks, but they are less often used for NLP tasks than RNNs because they don't take into account the temporal order of items in the input. However, they can be useful for document classification tasks. As you will recall from earlier chapters, the representations that are often used in classification depend only on the words that occur in the document—BoW and TF-IDF, for example—so, effective classification can often be accomplished without taking word order into account.

To classify documents with CNNs, we can represent a text as an array of vectors, where each word is mapped to a vector in a space made up of the full vocabulary. We can use word2vec, which we discussed in *Chapter 7*, to represent word vectors. Training a CNN for text classification with Keras is very similar to the training process that we worked through in MLP classification. We create a sequential model as we did earlier, but we add new convolutional layers and pooling layers.

We will not cover the details of using CNNs for classification, but they are another option for NLP classification. As in the case of RNNs, there are many available resources on this topic, and a good starting point is *Wikipedia* (`https://en.wikipedia.org/wiki/Convolutional_neural_network`).

Summary

In this chapter, we have explored applications of NNs to document classification in NLP. We covered the basic concepts of NNs, reviewed a simple MLP, and applied it to a binary classification problem. We also provided some suggestions for improving performance by modifying hyperparameters and tuning. Finally, we discussed the more advanced types of NNs—RNNs and CNNs.

In *Chapter 11*, we will cover the currently best-performing techniques in NLP—transformers and pretrained models.

11

Machine Learning Part 3 – Transformers and Large Language Models

In this chapter, we will cover the currently best-performing techniques in **natural language processing (NLP)** – **transformers** and **pretrained models**. We will discuss the concepts behind transformers and include examples of using transformers and **large language models (LLMs)** for text classification. The code for this chapter will be based on the TensorFlow/Keras Python libraries and the cloud services provided by OpenAI.

The topics covered in this chapter are important because although transformers and LLMs are only a few years old, they have become state-of-the-art for many different types of NLP applications. In fact, LLM systems such as ChatGPT have been widely covered in the press and you have undoubtedly encountered references to them. You have probably even used their online interfaces. In this chapter, you will learn how to work with the technology behind these systems, which should be part of the toolkit of every NLP developer.

In this chapter, we will cover the following topics:

- Overview of transformers and large language models
- **Bidirectional Encoder Representations from Transformers (BERT)** and its variants
- Using BERT – a classification example
- Cloud-based LLMs

We'll start by listing the technical resources that we'll use to run the examples in this chapter.

Technical requirements

The code that we will go over in this chapter makes use of a number of open source software libraries and resources. We have used many of these in earlier chapters, but we will list them here for convenience:

- The Tensorflow machine learning libraries: `hub`, `text`, and `tf-models`
- The Python numerical package, NumPy
- The Matplotlib plotting and graphical package
- The IMDb movie reviews dataset
- scikit-learn's `sklearn.model_selection` to do the training, validation, and test split
- A BERT model from TensorFlow Hub: we're using this one – `'small_bert/bert_en_uncased_L-4_H-512_A-8'` – but you can use any other BERT model you like, bearing in mind that larger models might take a long time to train

Note that we have kept the models relatively small here so that they don't require an especially powerful computer. The examples in this chapter were tested on a Windows 10 machine with an Intel 3.4 GHz CPU and 16 GB of RAM, without a separate GPU. Of course, more computing resources will speed up your training runs and enable you to use larger models.

The next section provides a brief description of the transformer and LLM technology that we'll be using.

Overview of transformers and LLMs

Transformers and LLMs are currently the best-performing technologies for **natural language understanding (NLU)**. This does not mean that the approaches covered in earlier chapters are obsolete. Depending on the requirements of a specific NLP project, some of the simpler approaches may be more practical or cost-effective. In this chapter, you will get information about the more recent approaches that you can use to make that decision.

There is a great deal of information about the theoretical aspects of these techniques available on the internet, but here we will focus on applications and explore how these technologies can be applied to solving practical NLU problems.

As we saw in *Chapter 10*, **recurrent neural networks (RNNs)** have been a very effective approach in NLP because they don't assume that the elements of input, specifically words, are independent, and so are able to take into account sequences of input elements such as the order of words in sentences. As we have seen, RNNs keep the memory of earlier inputs by using previous outputs as inputs to later layers. However, with RNNs, the effect of earlier inputs on the current input diminishes quickly as processing proceeds through the sequence.

When longer documents are processed, because of the context-dependent nature of natural language, even very distant parts of the text can have a strong effect on the current input. In fact, in some cases, distant inputs can be more important than more recent parts of the input. But when the data is a long sequence, processing with an RNN means that the earlier information will not be able to have much impact on the later processing. Some initial attempts to address this issue include **long short-term memory** (**LSTM**), which allows the processor to maintain the state and includes forget gates, and **gated recurrent units** (**GRUs**), a new and relatively fast type of LSTM, which we will not cover in this book. Instead, we will focus on more recent approaches such as attention and transformers.

Introducing attention

Attention is a technique that allows a network to learn where to pay attention to the input.

Initially, attention was used primarily in machine translation. The processing was based on an encoder-decoder architecture where a sentence was first encoded into a vector and then decoded into the translation. In the original encoder-decoder idea, each input sentence was encoded into a fixed-length vector. It turned out that it was difficult to encode all of the information in a sentence into a fixed-length vector, especially a long sentence. This is because more distant words that were outside of the scope of the fixed-length vector were not able to influence the result.

Encoding the sentence into a *set* of vectors, one per word, removed this limitation.

As one of the early papers on attention states, *"The most important distinguishing feature of this approach from the basic encoder-decoder is that it does not attempt to encode a whole input sentence into a single fixed-length vector. Instead, it encodes the input sentence into a sequence of vectors and chooses a subset of these vectors adaptively while decoding the translation. This frees a neural translation model from having to squash all the information of a source sentence, regardless of its length, into a fixed-length vector."* (Bahdanau, D., Cho, K., & Bengio, Y. (2014). *Neural machine translation by jointly learning to align and translate.* arXiv preprint arXiv:1409.0473.)

For machine translation applications, it is necessary both to encode the input text and to decode the results into the new language in order to produce the translated text. In this chapter, we will simplify this task by using a classification example that uses just the encoding part of the attention architecture.

A more recent technical development has been to demonstrate that one component of the attention architecture, RNNs, was not needed in order to get good results. This new development is called **transformers**, which we will briefly mention in the next section, and then illustrate with an in-depth example.

Applying attention in transformers

Transformers are a development of the attention approach that dispenses with the RNN part of the original attention systems. Transformers were introduced in the 2017 paper *Attention is all you need* (Ashish Vaswani, et al., 2017. *Attention is all you need*. In the Proceedings of the 31st International Conference on Neural Information Processing Systems (NIPS'17). Curran Associates Inc., Red Hook, NY, USA, 6000-6010). The paper showed that good results can be achieved just with attention. Nearly all research on NLP learning models is now based on transformers.

A second important technical component of the recent dramatic increases in NLP performance is the idea of pretraining models based on large amounts of existing data and making them available to NLP developers. The next section talks about the advantages of this approach.

Leveraging existing data – LLMs or pre-trained models

So far, in this book, we've created our own text representations (vectors) from training data. In our examples so far, all of the information that the model has about the language is contained in the training data, which is a very small sample of the full language. But if models start out with general knowledge of a language, they can take advantage of vast amounts of training data that would be impractical for a single project. This is called the **pretraining** of a model. These pretrained models can be reused for many projects because they capture general information about a language. Once a pretrained model is available, it can be fine-tuned to specific applications by supplying additional data

The next section will introduce one of the best-known and most important pretrained transformer models, BERT.

BERT and its variants

As an example of an LLM technology based on transformers, we will demonstrate the use of BERT, a widely used state-of-the-art system. BERT is an open source NLP approach developed by Google that is the foundation of today's state-of-the-art NLP systems. The source code for BERT is available at `https://github.com/google-research/bert`.

BERT's key technical innovation is that the training is bidirectional, that is, taking both previous and later words in input into account. A second innovation is that BERT's pretraining uses a masked language model, where the system masks out a word in the training data and attempts to predict it.

BERT also uses only the encoder part of the encoder-decoder architecture because, unlike machine translation systems, it focuses only on understanding; it doesn't produce language.

Another advantage of BERT, unlike the systems we've discussed earlier in this book, is that the training process is unsupervised. That is, the text that it is trained on does not need to be annotated or assigned any meaning by a human. Because it is unsupervised, the training process can take advantage of the enormous quantities of text available on the web, without needing to go through the expensive process of having humans review it and decide what it means.

The initial BERT system was published in 2018. Since then, the ideas behind BERT have been explored and expanded into many different variants. The different variants have various features that make them appropriate for addressing different requirements. Some of these features include faster training times, smaller models, or higher accuracy. *Table 11.1* shows a few of the common BERT variants and their specific features. Our example will use the original BERT system since it is the basis of all the other BERT versions:

Acronym	Name	Date	Features
BERT	Bidirectional Encoder Representations from Transformer	2018	The original BERT system.
BERT-Base			A number of models released by the original BERT authors.
RoBERTa	Robustly Optimized BERT pre-training approach	2019	In this approach, different parts of the sentences are masked in different epochs, which makes it more robust to variations in the training data.
ALBERT	A Lite BERT	2019	A version of BERT that shares parameters between layers in order to reduce the size of models.
DistilBERT		2020	Smaller and faster than BERT with good performance
TinyBERT		2019	Smaller and faster than BERT-Base with good performance; good for resource-restricted devices.

Table 11.1 – BERT variations

The next section will go through a hands-on example of a BERT application.

Using BERT – a classification example

In this example, we'll use BERT for classification, using the movie review dataset we saw in earlier chapters. We will start with a pretrained BERT model and *fine-tune* it to classify movie reviews. This is a process that you can follow if you want to apply BERT to your own data.

Using BERT for specific applications starts with one of the pretrained models available from TensorFlow Hub (`https://tfhub.dev/tensorflow`) and then fine-tuning it with training data that is specific to the application. It is recommended to start with one of the small BERT models, which have the same architecture as BERT but are faster to train. Generally, the smaller models are less accurate, but if their accuracy is adequate for the application, it isn't necessary to take the extra time and computer resources that would be needed to use a larger model. There are many models of various sizes that can be downloaded from TensorFlow Hub.

BERT models can also be cased or uncased, depending on whether they take the case of text into account. Uncased models will typically provide better results unless the application is one where the case of the text is informative, such as **named entity recognition (NER)**, where proper names are important.

In this example, we will work with the `small_bert/bert_en_uncased_L-4_H-512_A-8/1` model. It has the following properties, which are encoded in its name:

- Small BERT
- Uncased
- 4 hidden layers (L-4)
- A hidden size of 512
- 8 attention heads (A-8)

This model was trained on Wikipedia and BooksCorpus. This is a very large amount of text, but there are many pretrained models that were trained on much larger amounts of text, which we will discuss later in the chapter. Indeed, an important trend in NLP is developing and publishing models trained on larger and larger amounts of text.

The example that will be reviewed here is adapted from the TensorFlow tutorial for text classification with BERT. The full tutorial can be found here:

`https://colab.research.google.com/github/tensorflow/text/blob/master/docs/tutorials/classify_text_with_bert.ipynb#scrollTo=EqL7ihkN_862`

We'll start by installing and loading some basic libraries. We will be using a Jupyter notebook (you will recall that the process of setting up a Jupyter notebook was covered in detail in *Chapter 4*, and you can refer to *Chapter 4* for additional details if necessary):

```
!pip install -q -U "tensorflow-text==2.8.*"
!pip install -q tf-models-official==2.7.0
!pip install numpy==1.21
import os
import shutil
```

```
import tensorflow as tf
import tensorflow_hub as hub
import tensorflow_text as text
from official.nlp import optimization  # to create AdamW optimizer
import matplotlib.pyplot as plt #for plotting results
tf.get_logger().setLevel('ERROR')
```

Our BERT fine-tuned model will be developed through the following steps:

1. Installing data.

2. Splitting the data into training, validation, and testing subsets.

3. Loading a BERT model from TensorFlow Hub.

4. Building a model by combining BERT with a classifier.

5. Fine-tuning BERT to create a model.

6. Defining the loss function and metrics.

7. Defining the optimizer and number of training epochs.

8. Compiling the model.

9. Training the model.

10. Plotting the results of the training steps over the training epochs.

11. Evaluating the model with the test data.

12. Saving the model and using it to classify texts.

The following sections will go over each of these steps in detail.

Installing the data

The first step is to install the data. We will use the NLTK movie review dataset that we installed in *Chapter 10*. We will use the `tf.keras.utils.text_dataset_from_directory` utility to make a TensorFlow dataset from the movie review directory:

```
batch_size = 32
import matplotlib.pyplot as plt
tf.get_logger().setLevel('ERROR')
AUTOTUNE = tf.data.AUTOTUNE
raw_ds = tf.keras.utils.text_dataset_from_directory(
    './movie_reviews',
class_names = raw_ds.class_names
print(class_names)
```

There are 2,000 files in the dataset, divided into two classes, neg and pos. We print the class names in the final step as a check to make sure the class names are as expected. These steps can be used for any dataset that is contained in a directory structure with examples of the different classes contained in different directories with the class names as directory names.

Splitting the data into training, validation, and testing sets

The next step is to split the dataset into training, validation, and testing sets. As you will recall from earlier chapters, the training set is used to develop the model. The validation set, which is kept separate from the training set, is used to look at the performance of the system on data that it hasn't been trained on during the training process. In our example, we will use a common split of 80% training 10% for validation, and 10% for testing. The validation set can be used at the end of every training epoch, to see how training is progressing. The testing set is only used once, as a final evaluation:

```
from sklearn.model_selection import train_test_split
def partition_dataset_tf(dataset, ds_size, train_split=0.8, val_
split=0.1, test_split=0.1, shuffle=True, shuffle_size=1000):
    assert (train_split + test_split + val_split) == 1
    if shuffle:
        # Specify seed maintain the same split distribution between
runs for reproducibilty
        dataset = dataset.shuffle(shuffle_size, seed=42)
    train_size = int(train_split * ds_size)
    val_size = int(val_split * ds_size)
    train_ds = dataset.take(train_size)
    val_ds = dataset.skip(train_size).take(val_size)
    test_ds = dataset.skip(train_size).skip(val_size)
    return train_ds, val_ds, test_ds
train_ds,val_ds,test_ds = partition_dataset_tf(
    raw_ds,len(raw_ds))
```

Loading the BERT model

The next step is to load the BERT model we will fine-tune in this example, as shown in the following code block. As discussed previously, there are many BERT models to select from, but this model is a good choice to start with.

We will also need to provide a preprocessor to transform the text inputs into numeric token IDs before their input to BERT. We can use the matching preprocessor provided by TensorFlow for this model:

```
bert_model_name = 'small_bert/bert_en_uncased_L-4_H-512_A-8'
map_name_to_handle = {
    'small_bert/bert_en_uncased_L-4_H-512_A-8':
```

```
            'https://tfhub.dev/tensorflow/small_bert/bert_en_uncased_L-
4_H-512_A-8/1',
    }
map_model_to_preprocess = {
    'small_bert/bert_en_uncased_L-4_H-512_A-8':
        'https://tfhub.dev/tensorflow/bert_en_uncased_preprocess/3',
    }
tfhub_handle_encoder = map_name_to_handle[bert_model_name]
tfhub_handle_preprocess = map_model_to_preprocess[
    bert_model_name]
bert_preprocess_model = hub.KerasLayer(
    tfhub_handle_preprocess)
```

The code here specifies the model we'll use and defines some convenience variables to simplify reference to the model, the encoder, and the preprocessor.

Defining the model for fine-tuning

The following code defines the model we will use. We can increase the size of the parameter to the Dropout layer if desired to make the model robust to variations in the training data:

```
def build_classifier_model():
    text_input = tf.keras.layers.Input(shape=(),
        dtype=tf.string, name='text')
    preprocessing_layer = hub.KerasLayer(
        tfhub_handle_preprocess, name='preprocessing')
    encoder_inputs = preprocessing_layer(text_input)
    encoder = hub.KerasLayer(tfhub_handle_encoder,
        trainable = True, name='BERT_encoder')
    outputs = encoder(encoder_inputs)
    net = outputs['pooled_output']
    net = tf.keras.layers.Dropout(0.1)(net)
    net = tf.keras.layers.Dense(1, activation=None,
        name='classifier')(net)
    return tf.keras.Model(text_input, net)
# plot the model's structure as a check
tf.keras.utils.plot_model(classifier_model)
```

In *Figure 11.1*, we can see a visualization of the model's layers, including the text input layer, the preprocessing layer, the BERT layer, the dropout layer, and the final classifier layer. The visualization was produced by the last line in the code block. This structure corresponds to the structure we defined in the preceding code:

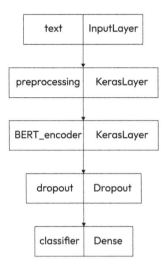

Figure 11.1 – Visualizing the model structure

Sanity checks such as this visualization are useful because, with larger datasets and models, the training process can be very lengthy, and if the structure of the model is not what was intended, a lot of time can be wasted in trying to train an incorrect model.

Defining the loss function and metrics

We will use a cross-entropy function for the loss function. **Cross-entropy** estimates the loss by scoring the average difference between the actual and predicted probability distributions for all classes. Since this is a binary classification problem (that is, there are only two outcomes, *positive* and *negative*), we'll use the `losses.BinaryCrossEntropy` loss function:

```
loss = tf.keras.losses.BinaryCrossentropy(from_logits=True)
metrics = tf.metrics.BinaryAccuracy()
```

A classification application with several possible outcomes, such as an intent identification problem where we have to decide which of 10 intents to assign to an input, would use categorical cross-entropy. Similarly, since this is a binary classification problem, the metric should be `binary accuracy`, rather than simply `accuracy`, which would be appropriate for a multi-class classification problem.

Defining the optimizer and the number of epochs

The optimizer improves the efficiency of the learning process. We're using the popular Adam optimizer here, and starting it off with a very small learning rate (`3e-5`), which is recommended for BERT. The optimizer will dynamically adjust the learning rate during training:

```
epochs = 15
```

```
steps_per_epoch = tf.data.experimental.cardinality(
    train_ds).numpy()
print(steps_per_epoch)

num_train_steps = steps_per_epoch * epochs
# a linear warmup phase over the first 10%
num_warmup_steps = int(0.1*num_train_steps)

init_lr = 3e-5
optimizer = optimization.create_optimizer(
        init_lr=init_lr, num_train_steps = num_train_steps,
        num_warmup_steps=num_warmup_steps,
        optimizer_type='adamw')
```

Note that we have selected 15 epochs of training. For the first training run, we'll try to balance the goals of training on enough epochs to get an accurate model and wasting time training on more epochs than needed. Once we get our results from the first training run, we can adjust the number of epochs to balance these goals.

Compiling the model

Using the classifier model in the call to `def build_classifier_model()`, we can compile the model with the loss, metrics, and optimizer, and take a look at the summary. It's a good idea to check the model before starting a lengthy training process to make sure the model looks as expected:

```
classifier_model.compile(optimizer=optimizer,
                         loss=loss,
                         metrics=metrics)
classifier_model.summary()
```

The summary of the model will look something like the following (we will only show a few lines because it is fairly long):

```
Model: model

_____

 Layer (type)                   Output Shape        Param
 #      Connected to
=====================================================================
=============================
 text (InputLayer)              [(None,)]              0            []

 preprocessing (KerasLayer)     {'input_mask':
(Non   0              ['text[0][0]']
                                 e, 128),
                                 'input_type_ids':
                                 (None, 128),
```

```
                                           'input_word_ids':
                                           (None, 128)}
```

The output here just summarizes the first two layers – input and preprocessing.

The next step is training the model.

Training the model

In the following code, we start the training process with a call to `classifier_model.fit(_)`. We supply this method with parameters for the training data, the validation data, the verbosity level, and the number of epochs (which we set earlier), as shown in this code:

```
print(f'Training model with {tfhub_handle_encoder}')

history = classifier_model.fit(x=train_ds,
                               validation_data=val_ds,
                               verbose = 2,
                               epochs=epochs)
Training model with https://tfhub.dev/tensorflow/small_bert/bert_en_
uncased_L-4_H-512_A-8/1
Epoch 1/15
50/50 - 189s - loss: 0.7015 - binary_accuracy: 0.5429 - val_loss:
0.6651 - val_binary_accuracy: 0.5365 - 189s/epoch - 4s/step
```

Note that the `classifier_model.fit()` method returns a `history` object, which will include information about the progress of the complete training process. We will use the `history` object to create plots of the training process. These will provide quite a bit of insight into what happened during training, and we will use this information to guide our next steps. We will see these plots in the next section.

Training times for transformer models can be quite lengthy. The time taken depends on the size of the dataset, the number of epochs, and the size of the model, but this example should probably not take more than an hour to train on a modern CPU. If running this example takes significantly longer than that, you may want to try testing with a higher verbosity level (2 is the maximum) so that you can get more information about what is going on in the training process.

At the end of this code block, we also see the results of processing the first epoch of training. We can see that the first epoch of training took 189 seconds. The loss was 0.7 and the accuracy was 0.54. The loss and accuracy after one epoch of training are not very good, but they will improve dramatically as training proceeds. In the next section, we will see how to show the training progress graphically.

Plotting the training process

After training is complete, we will want to see how the system's performance changes over training epochs. We can see this with the following code:

```
import matplotlib.pyplot as plt
!matplotlib inline
history_dict = history.history
print(history_dict.keys())

acc = history_dict['binary_accuracy']
val_acc = history_dict['val_binary_accuracy']
loss = history_dict['loss']
val_loss = history_dict['val_loss']

epochs = range(1, len(acc) + 1)
```

The preceding code defines some variables and gets the relevant metrics (`binary_accuracy` and `loss` for the training and validation data) from the model's `history` object. We are now ready to plot the progress of the training process. As usual, we will use Matplotlib to create our plots:

```
fig = plt.figure(figsize=(10, 6))
fig.tight_layout()
plt.subplot(2, 1, 1)
# r is for "solid red line"
plt.plot(epochs, loss, 'r', label='Training loss')
# b is for "solid blue line"
plt.plot(epochs, val_loss, 'b', label='Validation loss')
plt.title('Training and validation loss')
# plt.xlabel('Epochs')
plt.ylabel('Loss')
plt.legend()
plt.subplot(2, 1, 2)
plt.plot(epochs, acc, 'r', label='Training acc')
plt.plot(epochs, val_acc, 'b', label='Validation acc')
plt.title('Training and validation accuracy')
plt.xlabel('Epochs')
plt.ylabel('Accuracy')
plt.legend(loc='lower right')
plt.show()
dict_keys(['loss', 'binary_accuracy', 'val_loss',
    'val_binary_accuracy'])
```

In *Figure 11.2*, we see the plot of the decreasing loss and increasing accuracy over time as the model is trained. The dashed lines represent the training loss and accuracy, and the solid lines represent the validation loss and accuracy:

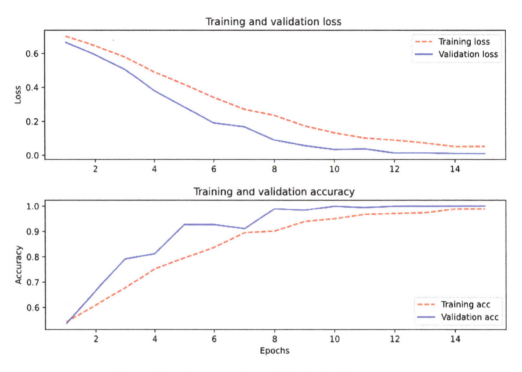

Figure 11.2 – Accuracy and loss during the training process

It is most typical for the validation accuracy to be less than the training accuracy, and for the validation loss to be greater than the training loss, but this will not necessarily be the case, depending on how the data is split between validation and training subsets. In this example, the validation loss is uniformly lower than the training loss and the validation accuracy is uniformly higher than the training accuracy. We can see from this plot that the system isn't changing after the first fourteen epochs. In fact, its performance is almost perfect.

Consequently, it is clear that there isn't any reason to train the system after this point. In comparison, look at the plots around epoch 4. We can see that it would not be a good idea to stop training after four epochs because loss is still decreasing and accuracy is still increasing. Another interesting observation that we can see in *Figure 11.2* around epoch 7 is that the accuracy seems to decrease a bit. If we had stopped training at epoch 7, we couldn't tell that accuracy would start to increase again at epoch 8. For that reason, it's a good idea to keep training until we either see the metrics level off or start to get consistently worse.

Now we have a trained model, and we'd like to see how it performs on previously unseen data. This unseen data is the test data that we set aside during the training, validation, and testing split.

Evaluating the model on the test data

After the training is complete, we can see how the model performs on the test data. This can be seen in the following output, where we can see that the system is doing very well. The accuracy is nearly 100% and the loss is near zero:

```
loss, accuracy = classifier_model.evaluate(test_ds)

print(f'Loss: {loss}')
print(f'Binary Accuracy: {accuracy}')

1/7 [===>..........................] - ETA: 9s - loss: 0.0239 -
binary_accuracy: 0.9688
2/7 [=======>......................] - ETA: 5s - loss: 0.0189 -
binary_accuracy: 0.9844
3/7 [===========>..................] - ETA: 4s - loss: 0.0163 -
binary_accuracy: 0.9896
4/7 [================>.............] - ETA: 3s - loss: 0.0140 -
binary_accuracy: 0.9922
5/7 [====================>.........] - ETA: 2s - loss: 0.0135 -
binary_accuracy: 0.9937
6/7 [========================>.....] - ETA: 1s - loss: 0.0134 -
binary_accuracy: 0.9948
7/7 [=============================] - ETA: 0s - loss: 0.0127 -
binary_accuracy: 0.9955
7/7 [=============================] - 8s 1s/step - loss: 0.0127 -
binary_accuracy: 0.9955
Loss: 0.012707981280982494
Accuracy: 0.9955357313156128
```

This is consistent with the system performance during training that we saw in *Figure 11.2*.

It looks like we have a very accurate model. If we want to use it later on, we can save it.

Saving the model for inference

The final step is to save the fine-tuned model for later use – for example, if the model is to be used in a production system, or if we want to use it in further experiments. The code for saving the model can be seen here:

```
dataset_name = 'movie_reviews'
saved_model_path = './{}_bert'.format(dataset_name.replace('/', '_'))

classifier_model.save(saved_model_path, include_optimizer=False)
```

```
reloaded_model = tf.saved_model.load(saved_model_path)
]
```

In the code here, we show both saving the model and then reloading it from the saved location.

As we saw in this section, BERT can be trained to achieve very good performance by fine-tuning it with a relatively small (2,000-item) dataset. This makes it a good choice for many practical problems. Looking back at the example of classification with the multi-layer perceptron in *Chapter 10*, we saw that the accuracy (as shown in *Figure 10.4*) was never better than about 80% for the validation data, even after 20 epochs of training. Clearly, BERT does much better than that.

Although BERT is an excellent system, it has recently been surpassed by very large cloud-based pretrained LLMs. We will describe them in the next section.

Cloud-based LLMs

Recently, there have been a number of cloud-based pretrained large language models that have shown very impressive performance because they have been trained on very large amounts of data. In contrast to BERT, they are too large to be downloaded and used locally. In addition, some are closed and proprietary and can't be downloaded for that reason. These newer models are based on the same principles as BERT, and they have shown a very impressive performance. This impressive performance is due to the fact that these models have been trained with much larger amounts of data than BERT. Because they cannot be downloaded, it is important to keep in mind that they aren't appropriate for every application. Specifically, if there are any privacy or security concerns regarding the data, it may not be a good idea to send it to the cloud for processing. Some of these systems are GPT-2, GPT-3, GPT-4, ChatGPT, and OPT-175B, and new LLMs are being published on a frequent basis.

The recent dramatic advances in NLP represented by these systems are made possible by three related technical advances. One is the development of techniques such as attention, which are much more able to capture relationships among words in texts than previous approaches such as RNNs, and which scale much better than the rule-based approaches that we covered in *Chapter 8*. The second factor is the availability of massive amounts of training data, primarily in the form of text data on the World Wide Web. The third factor is the tremendous increase in computer resources available for processing this data and training LLMs.

So far in the systems we've discussed, all of the knowledge of a language that goes into the creation of a model for a specific application is derived from the training data. The process starts without knowing anything about the language. LLMs, on the other hand, come with models that have been *pretrained* through processing very large amounts of more or less generic text, and as a consequence have a basic foundation of information about the language. Additional training data can be used for *fine-tuning* the model so that it can handle inputs that are specific to the application. An important aspect of fine-tuning a model for a specific application is to minimize the amount of new data that is needed for fine-tuning. This is a cutting-edge area in NLP research and you may find references to training approaches called **few-shot learning**, which is learning to recognize a new class with only a few examples, or even **zero-shot learning**, which enables a system to identify a class without having seen any examples of that class in the training data.

In the next section, we'll take a look at one of the currently most popular LLMs, ChatGPT.

ChatGPT

ChatGPT (`https://openai.com/blog/chatgpt/`) is a system that can interact with users about generic information in a very capable way. Although at the time of writing, it is hard to customize ChatGPT for specific applications, it can be useful for other purposes than customized natural language applications. For example, it can very easily be used to generate training data for a conventional application. If we wanted to develop a banking application using some of the techniques discussed earlier in this book, we would need training data to provide the system with examples of how users might ask the system questions. Typically, this involves a process of collecting actual user input, which could be very time-consuming. ChatGPT could be used to generate training data instead, by simply asking it for examples. For example, for the prompt *give me 10 examples of how someone might ask for their checking balance*, ChatGPT responded with the sentences in *Figure 11.3*:

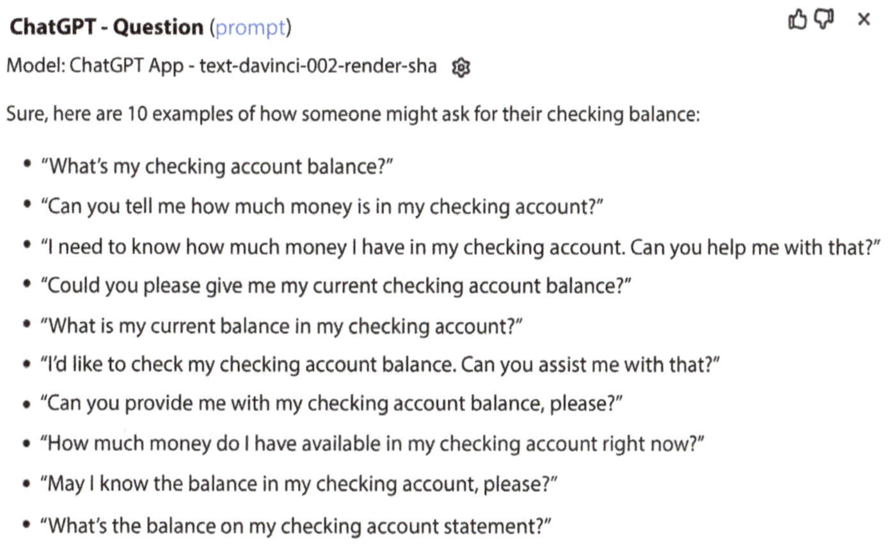

ChatGPT - Question (prompt)

Model: ChatGPT App - text-davinci-002-render-sha

Sure, here are 10 examples of how someone might ask for their checking balance:

- "What's my checking account balance?"
- "Can you tell me how much money is in my checking account?"
- "I need to know how much money I have in my checking account. Can you help me with that?"
- "Could you please give me my current checking account balance?"
- "What is my current balance in my checking account?"
- "I'd like to check my checking account balance. Can you assist me with that?"
- "Can you provide me with my checking account balance, please?"
- "How much money do I have available in my checking account right now?"
- "May I know the balance in my checking account, please?"
- "What's the balance on my checking account statement?"

Figure 11.3 – GPT-3 generated training data for a banking application

Most of these seem like pretty reasonable queries about a checking account, but some of them don't seem very natural. For that reason, data generated in this way always needs to be reviewed. For example, a developer might decide not to include the second to the last example in a training set because it sounds stilted, but overall, this technique has the potential to save developers quite a bit of time.

Applying GPT-3

Another well-known LLM, GPT-3, can also be fine-tuned with application-specific data, which should result in better performance. To do this, you need an OpenAI key because using GPT-3 is a paid service. Both fine-tuning to prepare the model and using the fine-tuned model to process new data at inference time will incur a cost, so it is important to verify that the training process is performing as expected before training with a large dataset and incurring the associated expense.

OpenAI recommends the following steps to fine-tune a GPT-3 model.

1. Sign up for an account at https://openai.com/ and obtain an API key. The API key will be used to track your usage and charge your account accordingly.

2. Install the OpenAI **command-line interface (CLI)** with the following command:

    ```
    ! pip install --upgrade openai
    ```

This command can be used at a terminal prompt in Unix-like systems (some developers have reported problems with Windows or macOS). Alternatively, you can install GPT-3 to be used in a Jupyter notebook with the following code:

```
!pip install --upgrade openai
```

All of the following examples assume that the code is running in a Jupyter notebook:

1. Set your API key:

    ```
    api_key =<your API key>
    openai.api_key = api_key
    ```

2. The next step is to specify the training data that you will use for fine-tuning GPT-3 for your application. This is very similar to the process of training any NLP system; however, GPT-3 has a specific format that must be used for training data. This format uses a syntax called JSONL, where every line is an independent JSON expression. For example, if we want to fine-tune GPT-3 to classify movie reviews, a couple of data items would look like the following (omitting some of the text for clarity):

    ```
    {"prompt":"this film is extraordinarily horrendous and i'm
    not going to waste any more words on it . ","completion":"
    negative"}
    {"prompt":"9 : its pathetic attempt at \" improving \" on a
    shakespeare classic . 8 : its just another piece of teen fluff .
    7 : kids in high school are not that witty . … ","completion":"
    negative"}
    {"prompt":"claire danes , giovanni ribisi , and omar epps make a
    likable trio of protagonists , …","completion":" negative"}
    ```

Each item consists of a JSON dict with two keys, `prompt` and `completion`. `prompt` is the text to be classified, and `completion` is the correct classification. All three of these items are negative reviews, so the completions are all marked as `negative`.

It might not always be convenient to get your data into this format if it is already in another format, but OpenAI provides a useful tool for converting other formats into JSONL. It accepts a wide range of input formats, such as CSV, TSV, XLSX, and JSON, with the only requirement for the input being that it contains two columns with `prompt` and `completion` headers. *Table 11.2* shows a few cells from an Excel spreadsheet with some movie reviews as an example:

prompt	completion
kolya is one of the richest films i've seen in some time . zdenek sverak plays a confirmed old bachelor (who's likely to remain so) , who finds his life as a czech cellist increasingly impacted by the five-year old boy that he's taking care of …	positive
this three hour movie opens up with a view of singer/guitar player/musician/composer frank zappa rehearsing with his fellow band members . all the rest displays a compilation of footage , mostly from the concert at the palladium in new york city , halloween 1979 …	positive
`strange days' chronicles the last two days of 1999 in los angeles . as the locals gear up for the new millenium , lenny nero (ralph fiennes) goes about his business …	positive

Table 11.2 – Movie review data for fine-tuning GPT-3

To convert one of these alternative formats into JSONL, you can use the `fine_tunes.prepare_data` tool, as shown here, assuming that your data is contained in the `movies.csv` file:

```
!openai tools fine_tunes.prepare_data -f ./movies.csv -q
```

The `fine_tunes.prepare_data` utility will create a JSONL file of the data and will also provide some diagnostic information that can help improve the data. The most important diagnostic that it provides is whether or not the amount of data is sufficient. OpenAI recommends several hundred examples of good performance. Other diagnostics include various types of formatting information such as separators between the prompts and the completions.

After the data is correctly formatted, you can upload it to your OpenAI account and save the filename:

```
file_name = "./movies_prepared.jsonl"
upload_response = openai.File.create(
    file=open(file_name, "rb"),
    purpose='fine-tune'
```

```
)
file_id = upload_response.id
```

The next step is to create and save a fine-tuned model. There are several different OpenAI models that can be used. The one we're using here, `ada`, is the fastest and least expensive, and does a good job on many classification tasks:

```
openai.FineTune.create(training_file=file_id, model="ada")
fine_tuned_model = fine_tune_response.fine_tuned_model
```

Finally, we can test the model with a new prompt:

```
answer = openai.Completion.create(
   model = fine_tuned_model,
     engine = "ada",
   prompt = " I don't like this movie ",
   max_tokens = 10, # Change amount of tokens for longer completion
   temperature = 0
)
answer['choices'][0]['text']
```

In this example, since we are only using a few fine-tuning utterances, the results will not be very good. You are encouraged to experiment with larger amounts of training data.

Summary

This chapter covered the currently best-performing techniques in NLP – transformers and pretrained models. In addition, we have demonstrated how they can be applied to processing your own application-specific data, using both local pretrained models and cloud-based models.

Specifically, you learned about the basic concepts behind attention, transformers, and pretrained models, and then applied the BERT pretrained transformer system to a classification problem. Finally, we looked at using the cloud-based GPT-3 systems for generating data and for processing application-specific data.

In *Chapter 12*, we will turn to a different topic – unsupervised learning. Up to this point, all of our models have been *supervised*, which you will recall means that the data has been annotated with the correct processing result. Next, we will discuss applications of *unsupervised* learning. These applications include topic modeling and clustering. We will also talk about the value of unsupervised learning for exploratory applications and maximizing scarce data. It will also address types of partial supervision, including weak supervision and distant supervision.

12
Applying Unsupervised Learning Approaches

In earlier chapters, such as *Chapter 5*, we discussed the fact that supervised learning requires annotated data, where a human annotator makes a decision about how a **natural language processing** (NLP) system should analyze it – that is, a human has *annotated* it. For example, with the movie review data, a human has looked at each review and decided whether it is positive or negative. We also pointed out that this annotation process can be expensive and time-consuming.

In this chapter, we will look at techniques that don't require annotated data, thereby saving this time-consuming step in data preparation. Although unsupervised learning will not be suitable for every NLP problem, it is very useful to have an understanding of the general area so that you can decide how to incorporate it into your NLP projects.

At a deeper level, we will discuss applications of unsupervised learning, such as topic modeling, including the value of unsupervised learning for exploratory applications and maximizing scarce data. We will also cover label generation in unsupervised classification and mention some approaches to make the most of limited labeled data, with techniques for partial supervision.

In this chapter, we will cover the following topics:

- What is unsupervised learning?
- Topic modeling using clustering techniques and label derivation
- Making the most of data with partial supervision

What is unsupervised learning?

The applications that we worked with in earlier chapters were based on data that was manually categorized by human annotators. For example, each review in the movie review corpus that we have used several times was read by a human annotator and assigned a category, *positive* or *negative*, based on the human's opinion. The review-category pairs were then used to train models, using the machine learning algorithms that we previously learned about to categorize new reviews. This whole process is called **supervised learning** because the training process is, in effect, *supervised* by the training data. The training data labeled by humans is referred to as the *gold standard* or *ground truth*.

Supervised approaches have some disadvantages, however. The most obvious disadvantage is the cost of developing the ground-truth data because of the cost of human annotators. Another consideration is the possibility that the manual annotations from different annotators, or even the same annotator at different times, will be inconsistent. Inconsistent annotation can also occur if the data labels themselves are subjective or not clear-cut, which makes it harder for the annotators to agree on the correct annotation.

For many applications, supervised approaches are the only option, but there are other applications, which we will be exploring in this chapter, where **unsupervised techniques** are useful.

These unsupervised applications don't require labeled training data because what we want to learn from the natural language data doesn't require any human judgment. Rather, it can be found by just examining the raw text, which can be done by an algorithm. These kinds of applications include grouping documents by similarity and computing the similarity of documents. In particular, we will be looking at **clustering**, which is the process of putting data, documents in particular, into similar groups. Finding similar groups of documents is often a first step in the process of developing a classification application, before the categories of documents are known. Once the clusters are identified, there are additional techniques that can help find human-readable labels for the clusters, although in some cases, it is easy to identify how the clusters should be labeled by manual inspection of the clusters. We will look at tools to find cluster labels later in this chapter.

In addition to clustering, another important application of unsupervised learning is the training process for the **large language models (LLMs)** that we covered in *Chapter 11*. Training LLMs doesn't require any supervision because the training process only looks at words in the context of other words. However, we will not cover the training process for LLMs in this book because this is a very computationally intensive process and requires expensive computational resources that are not available to the vast majority of developers. In addition, LLM training is not needed for most practical applications, since existing LLMs that have already been trained are widely available.

In this chapter, we will illustrate in detail a practical NLP problem where unsupervised learning can be very useful – *topic modeling*. In topic modeling, we start with a collection of text items, such as documents or user inputs in chatbot applications, but we don't have a pre-determined set of categories. Instead, we use the words in the texts themselves to find semantic similarities among the texts that enable us to group them into categories, or topics.

We will start by reviewing a few general considerations about semantically grouping similar texts.

Topic modeling using clustering techniques and label derivation

We'll start our exploration of topic modeling by looking at some considerations relating to grouping semantically similar documents in general, and then we'll look at a specific example.

Grouping semantically similar documents

Like most of the machine learning problems we've discussed so far, the overall task generally breaks down into two sub-problems, representing the data and performing a task based on the representations. We'll look at these two sub-problems next.

Representing the data

The data representations we've looked at so far were reviewed in *Chapter 7*. These approaches included the simple **bag of words (BoW)** variants, **term frequency - inverse document frequency (TF-IDF)**, and newer approaches, including **Word2Vec**. Word2Vec is based on word vectors, which are vectors that represent words in isolation, without taking into account the context in which they occur. A newer representation, used in the **BERT** system that we discussed in the previous chapter, takes into account the contexts of words in a sentence or document to create numerical word representations, or *embeddings*. We will use BERT embeddings in this chapter to uncover similarities among documents.

Working with data representations

This section will discuss two aspects of processing embeddings. First, we will discuss grouping similar texts into clusters, and then we will discuss visualizing the clusters.

Clustering – grouping similar items

Clustering is the main NLP task we'll talk about in this chapter. Clustering is the name for a wide variety of algorithms that attempt to group data items together, based on similarities in the data representation. Clustering can be used with any dataset, whether or not the data items are text-based, as long as there is a numerical way of representing their similarity. In this chapter, we will review a practical set of tools to perform clustering, but you should be aware that there are many other options, and undoubtedly, there will be many new ones as technology moves forward. Two common approaches to clustering are k-means and HDBSCAN:

- **k-means**: The k-means algorithm, which we reviewed in *Chapter 6*, is a very common clustering approach in which data points are initially randomly assigned to one of k clusters, the means of the clusters are computed, and the distance of the data points from the centers of the clusters is minimized through an iterative process. The value of k, or the number of clusters, is a hyperparameter that is selected by the developer. It can be considered to be the most *useful* number of clusters for the application. The k-means algorithm is often used because it is efficient and easy to implement, but other clustering algorithms – in particular, HDBSCAN – can yield better results.

- **HDBSCAN**: HDBSCAN is another popular clustering algorithm and stands for **Hierarchical Density-Based Spatial Clustering of Applications with Noise**. HDBSCAN takes into account the density of the data points within the clusters. Because of this, it is able to find oddly sized and differently sized clusters. It can also detect outliers or items that don't fit well into a cluster, while k-means forces every item into a cluster.

Visualizing the clusters

Visualization is very important in unsupervised approaches, such as clustering, because it allows us to see what groups of similar data items look like and helps us judge whether or not the grouping result is useful. While clusters of similar items can be represented in any number of dimensions, we are only able to effectively visualize clusters in, at most, three dimensions. Consequently, in practice, dimensionality reduction is needed to reduce the number of dimensions. We will use a tool called **Uniform Manifold Approximation and Projection** (**UMAP**) for dimensionality reduction.

In the next section, we will use clustering and visualization to illustrate a specific application of unsupervised learning called topic modeling. The general problem that can be solved using topic modeling is that of classifying documents into different topics. The unique characteristic of this technique is that, unlike the classification examples we saw in earlier chapters, we don't know what the topics are at the outset. Topic modeling can help us identify groups of similar documents, even if we don't know what the eventual categories will be.

In this example, we will use BERT transformer embeddings to represent the documents and HDBSCAN for clustering. Specifically, we will use the BERTopic Python library found at `https://maartengr.github.io/BERTopic/index.html`. The BERTopic library is customizable, but we will stick with the default settings for the most part in our example.

The data we'll look at is a well-known dataset called `20 newsgroups`, which is a collection of 20,000 newsgroup documents from 20 different internet newsgroups. This is a popular dataset that is often used in text processing. The data is email messages of varying lengths that are posted to newsgroups. Here's one example of a short message from this dataset, with the email headers removed:

```
I searched the U Mich archives fairly thoroughly for 3D graphics
packages,
I always thought it to be a mirror of sumex-aim.stanford.edu... I was
wrong.
I'll look into GrafSys... it does sound interesting!
```

```
Thanks Cheinan.
BobC
```

The 20 newsgroups dataset can be imported from the scikit-learn datasets or downloaded from the following website: http://qwone.com/~jason/20Newsgroups/.

> **Dataset citation**
>
> Ken Lang, *Newsweeder: Learning to filter netnews*, 1995, *Proceedings of the Twelfth International Conference on Machine Learning*, 331–339

In the following sections, we will go over topic modeling in detail using the 20 newsgroups dataset and the BERTopic package. We will create embeddings, construct the model, generate proposed labels for the topics, and visualize the resulting clusters. The last step will be to show the process of finding topics for new documents using our model.

Applying BERTopic to 20 newsgroups

The first step in this application is to install BERTopic and import the necessary libraries in a Jupyter notebook, as shown here:

```
!pip install bertopic
from sklearn.datasets import fetch_20newsgroups
from sklearn.feature_extraction.text import CountVectorizer
from sentence_transformers import SentenceTransformer
from bertopic import BERTopic
from umap import UMAP
from hdbscan import HDBSCAN
# install data
docs = fetch_20newsgroups(subset='all', remove=('headers', 'footers',
'quotes'))['data']
```

Embeddings

The next step is to prepare the data representations, or embeddings, shown in the following code block. Because this is a slow process, it is useful to set show_progress_bar () to True so that we can ensure that the process is moving along, even if it takes a long time:

```
# Prepare embeddings
docs = fetch_20newsgroups(subset='all', remove=('headers', 'footers',
'quotes'))['data']
#The model is a Hugging Face transformer model
embedding_model = SentenceTransformer("all-MiniLM-L6-v2")
corpus_embeddings = embedding_model.encode(docs, show_progress_bar =
True)
```

```
Batches: 100%|####################################################
################| 589/589 [21:48<00:00,  2.22s/it]
```

Instead of the earlier BERT word embeddings that we worked with in *Chapter 11* in this exercise, we will work with sentence embeddings using **Sentence Bert (SBERT)**.

SBERT produces one embedding per sentence. We will use a package called `SentenceTransformers`, available from Hugging Face, and the `all-MiniLM-L6-v2` model, which is recommended by BERTopic. However, many other transformer models can be used. These include, for example, the `en_core_web_trf` spaCy model, or the `distilbert-base-cased` Hugging Face model. BERTopic provides a guide to a wide selection of other models that you can use at `https://maartengr.github.io/BERTopic/getting_started/embeddings/embeddings.html`.

We can see what the actual embeddings look like in the following output:

```
corpus_embeddings.view()
array([[ 0.002078  ,  0.02345043,  0.02480883, ...,  0.00143592,
         0.0151075 ,  0.05287581],
       [ 0.05006033,  0.02698092, -0.00886482, ..., -0.00887168,
        -0.06737082,  0.05656359],
       [ 0.01640477,  0.08100049, -0.04953594, ..., -0.04184629,
        -0.07800221, -0.03130952],
       ...,
       [-0.00509084,  0.01817271,  0.04388074, ...,  0.01331367,
        -0.05997065, -0.05430664],
       [ 0.03508159, -0.05842971, -0.03385153, ..., -0.02824297,
        -0.05223113,  0.03760364],
       [-0.06498063, -0.01133722,  0.03949645, ..., -0.03573753,
         0.07217913,  0.02192113]], dtype=float32)
```

Using the `corpus_embeddings.view()` method shown here, we can see a summary of the embeddings, which are an array of arrays of floating point numbers. Looking directly at the embeddings isn't particularly useful itself, but it gives you something of a sense of what the actual data looks like.

Constructing the BERTopic model

Once the embeddings have been computed, we can build the BERTopic model. The BERTopic model can take a large number of parameters, so we won't show them all. We will show some useful ones, but there are many more, and you can consult the BERTopic documentation for additional ideas. The BERTopic model can be constructed very simply, with just the documents and the embeddings as parameters, as shown here:

```
model = BERTopic().fit(docs, corpus_embeddings)
```

This simple model has defaults for the parameters that will usually lead to a reasonable result. However, to illustrate some of the flexibility that's possible with BERTopic, we'll show next how the model can be constructed with a richer set of parameters, with the code shown here:

```
from sklearn.feature_extraction.text import CountVectorizer
vectorizer_model = CountVectorizer(stop_words = "english", max_df =
.95, min_df = .01)
# setting parameters for HDBSCAN (clustering) and UMAP (dimensionality
reduction)
hdbscan_model = HDBSCAN(min_cluster_size = 30, metric = 'euclidean',
prediction_data = True)
umap_model = UMAP(n_neighbors = 15, n_components = 10, metric =
'cosine', low_memory = False)

# Train BERTopic
model = BERTopic(
    vectorizer_model = vectorizer_model,
    nr_topics = 'auto',
    top_n_words = 10,
    umap_model = umap_model,
    hdbscan_model = hdbscan_model,
    min_topic_size = 30,
    calculate_probabilities = True).fit(docs, corpus_embeddings)
```

In the preceding code, we start by defining several useful models. The first one is CountVectorizer, which we saw in *Chapter 7* and *Chapter 10*, where we used it to vectorize text documents to formats such as BoW. Here, we will use the vectorizer to remove stopwords after dimensionality reduction and clustering so that they don't end up being included in topic labels.

Stopwords should not be removed from documents before preparing embeddings because the transformers have been trained on normal text, including stopwords, and the models will be less effective without the stopwords.

The vectorizer model parameters indicate that the model is constructed with English stopwords and that only words that occur in fewer than 95% of the documents and more than 1% are included. This is to exclude extremely common and extremely rare words from the model, which are unlikely to be helpful to distinguish topics.

The second model we will define is the HDBSCAN model, which is used for clustering. Some of the parameters include the following:

- The min_cluster_size parameter is the minimum number of documents that we want to be in a cluster. Here, we've selected '30' as the minimum cluster size. This parameter can vary, depending on the problem you're trying to solve, but you might want to consider a larger number if the number of documents in the dataset is large.

- `prediction_data` is set to `True` if we want to be able to predict the topics of new documents after the model is trained.

The next model is the **uniform manifold approximation and projection (UMAP)** model, which is used for dimensionality reduction. Besides making it easier to visualize multidimensional data, UMAP also makes it easier to cluster. Some of its parameters include the following:

- `n-neighbors`: This constrains the size of the area that UMAP will look at when learning the structure of the data.

- `n-components`: This determines the dimensionality of the reduced dimension space we will use.

- `low_memory`: This determines system memory management behavior. If the parameter is `False`, the algorithm will use a faster and more memory-intensive approach. If running out of memory is a problem with large datasets, you can set `low_memory` to `True`.

With these preliminary models defined, we can move on to defining the BERTopic model itself. The parameters that are set in this example are the following:

- The models for the three tools we've already defined – the vectorizer model, the UMAP model, and the HDBSCAN model.

- `nr_topics`: If we have an idea of how many topics we want to find, this parameter can be set to that number. In this case, it is set to `auto`, and HDBSCAN will find a good estimate of the number of topics.

- `top_n_words`: When generating labels, BERTopic should consider only the n most frequent words in a cluster.

- `min_topic_size`: The minimum number of documents that should be considered to form a topic.

- `calculate_probabilities`: This calculates the probabilities of all topics per document.

Running the code in this section will create clusters from the original documents. The number and size of the clusters will vary, depending on the parameters that were set in the generation of the model. You are encouraged to try different settings of the parameters and compare the results. As you think about the goals of your application, consider whether some of the settings seem to yield more or less helpful results.

Up to this point, we have used clusters of similar documents, but they are not labeled with any categories. It would be more useful if the clusters had names. Let's see how we can get names or labels for the clusters that relate to what they are about. The next section will review ways to label the clusters.

Labeling

Once we have the clusters, there are various approaches to labeling them with topics. A manual approach, where you just look at the documents and think about what a good label might be, should not necessarily be ruled out. However, there are automatic approaches that can suggest topics based on the words that occur in the documents in the various clusters.

BERTopic uses a method to suggest topic labels called class-based `tf-idf`, or `c-tf-idf`. You will recall that TF-IDF was discussed in earlier chapters as a document vectorization approach that identifies the most diagnostic terms for a document class. It does this by taking into account the frequency of a term in a document, compared to its frequency in the overall document collection. Terms that are frequent in one document but not overall in the dataset are likely to be indicative that a document should be assigned to a specific category.

Class-based TF-IDF takes this intuition a step further by treating each cluster as if it were a single document. It then looks at the terms that occur frequently in a cluster, but not overall in the entire dataset, to identify words that are useful to label a topic. Using this metric, labels can be generated for the topics using the most diagnostic words for each cluster. We can see in *Table 12.1* the labels generated for the 10 most frequent topics in a dataset, along with the number of documents in each topic.

Note that the first topic in the table, *-1*, is a catchall topic that includes the documents that don't fall into any topic.

	Topic	Count	Name
0	-1	6,928	`-1_maxaxaxaxaxaxaxaxaxaxaxaxaxaxaxax_dont_know_like`
1	0	1,820	`0_game_team_games_players`
2	1	1,569	`1_space_launch_nasa_orbit`
3	2	1,313	`2_car_bike_engine_cars`
4	3	1,168	`3_image_jpeg_window_file`
5	4	990	`4_armenian_armenians_people_turkish`
6	5	662	`5_drive_scsi_drives_ide`
7	6	636	`6_key_encryption_clipper_chip`
8	7	633	`7_god_atheists_believe_atheism`
9	8	427	`8_cheek___`
10	9	423	`9_israel_israeli_jews_arab`

Table 12.1 – The top 10 topics and automatically generated labels

Visualization

Visualization is extremely useful in unsupervised learning, because deciding how to proceed in the development process often depends on intuitions that can be assisted by looking at results in different ways.

There are many ways of visualizing the results of topic modeling. One useful visualization can be obtained with a BERTopic method, `model.visualize_barchart()`, which shows the top topics and their top words, as shown in *Figure 12.1*. For example, looking at the top words in **Topic 1** suggests that this topic is about space, and *Table 12.1* suggests that this topic could be labeled `space_launch_nasa_orbit`:

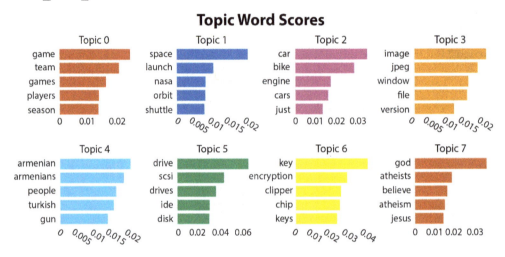

Figure 12.1 – The top seven topics and their most significant words

Another important visualization technique that is often used in clustering is representing each item in a cluster as a dot, representing their similarities by the distance between the dots, and assigning different colors or markers to the clusters. This clustering can be produced with a BERTopic method, `visualize_documents()`, as shown in the following code with the minimal `docs` and `embeddings` parameters:

```
model.visualize_documents(docs, embeddings = corpus_embeddings)
```

Additional parameters can be set to configure a cluster plot in various ways, as documented in the BERTopic documentation. For example, you can set the display to show or hide the cluster labels, or to only show the top topics.

The clustering for the top seven topics in the `20 newsgroups` data can be seen in *Figure 12.2*.

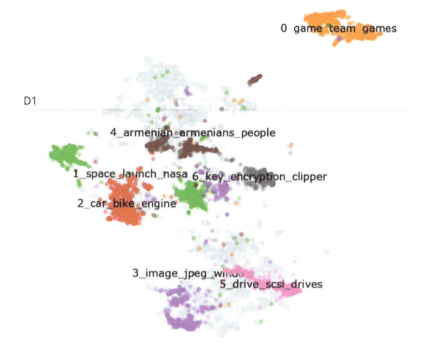

Figure 12.2 – The clustered documents for the top seven topics in the
20 newsgroups dataset with their generated labels

There are seven topics displayed in *Figure 12.2*, corresponding to the seven topics in *Figure 12.1*, each labeled with its automatically generated label. In addition, there are documents in *Figure 12.2* shown in light gray that were not assigned to any cluster. These are the documents shown as belonging to topic *-1* in *Table 12.1*.

One insight we can gain from the cluster display is the fact that there are many unassigned documents, which means that we may want to increase the number of topics to look for (by increasing the `nr_topics` parameter to `model()`). Another insight is that some of the topics – for example, `1_space_launch_nasa` – seem to be split over several clusters. This means that it might be more meaningful for them to be considered as separate topics. As in the case of unclassified documents, we can investigate this possibility by increasing the number of topics to look for.

We have shown two of the most useful BERTopic visualizations, but there are many more. You are encouraged to consult the BERTopic documentation for other ideas.

Finding the topic of a new document

Once the clustering is complete, the model can be used to find the topics of new documents, similar to a classification application. This can be done with the model.transform() method, as shown here:

```
sentence_model = SentenceTransformer("all-MiniLM-L6-v2")
new_docs = ["I'm looking for a new graphics card","when is the next
nasa launch"]
embeddings = sentence_model.encode(new_docs)
topics, probs = model.transform(new_docs,embeddings)
print(topics)
[-1, 3]
```

The predicted topics for the two documents in new_docs are -1 (no topic) and 3, which stands for 3_space_launch_orbit_nasa.

After clustering and topic labeling

Recall that the most common application of clustering is to explore the data, often in preparation to develop a supervised classification application. We can look at the clustering results and decide, for example, that some of the clusters are very similar and close to each other. In that case, it might be difficult to distinguish documents in those topics from each other, even after supervised training, and it would be a good idea to combine those clusters into a single topic. Similarly, if we find very small clusters, we would probably get more reliable results if the small clusters are combined with similar large clusters.

We can also modify labels to make them more helpful or informative. For example, topic *1* in *Table 12.1* is 1_space_launch_nasa_orbit. You might decide that space would be a simpler label that is, nevertheless, just as informative as the automatically generated label.

After making any adjustments to the clusters and labels that you find helpful, the result will be a supervised dataset, just like the supervised datasets we worked with, such as the movie reviews. You can use this as you would use any supervised dataset in NLP applications.

While unsupervised topic modeling can be a very useful technique, it is also possible to take advantage of data that is even partially supervised. We will summarize some of the ideas that the NLP research community has explored to use partially annotated data in the next section.

Making the most of data with weak supervision

In between completely supervised and unsupervised learning are several approaches to partial supervision, where only x data is supervised. Like unsupervised approaches, the goal of these techniques is to make the most of supervised data, which can be expensive to obtain. One advantage of partial supervision over unsupervised approaches is that unsupervised results don't automatically have useful labels. The labels have to be supplied, either manually or through some of the techniques we saw earlier in this chapter. In general, with weak supervision, the labels are supplied based on the subset of the data that is supervised.

This is an active research area, and we will not go into it in detail. However, it is useful to know what the general tactics of weak supervision are so that you will be able to apply them as they relate to specific tasks, depending on the kind of labeled data that is available.

Some tactics for weak supervision include the following:

- Incomplete supervision, where only partial data has ground-truth labels
- Inexact supervision, where the data has coarse-grained labels
- Inaccurate supervision, where some of the labels might be incorrect
- Semi-supervised learning, where some predefined labels are provided to help push a model toward known classes

These approaches are worth considering in applications where full annotation is too expensive or takes too long. In addition, they can also be useful in situations where unsupervised learning would be problematic because there are predefined labels that are required by other parts of the overall application, such as a database.

Summary

In this chapter, you learned about the basic concepts of unsupervised learning. We also worked through a specific application of unsupervised learning, topic modeling, using a BERT-based tool called BERTopic. We used the BERTopic package to identify clusters of semantically similar documents and propose labels for the clusters based on the words they contain, without needing to use any supervised annotations of the cluster topics.

In the next chapter, *Chapter 13*, we will address the question of measuring how good our results are using quantitative techniques. Quantitative evaluation is useful in research applications to compare results to those from previous research, and it is useful in practical applications to ensure that the techniques being used meet the application's requirements. Although evaluation was briefly discussed in earlier chapters, *Chapter 13* will discuss it in depth. It will include segmenting data into training, validation, and test data, evaluation with cross-validation, evaluation metrics such as precision and recall, the area under the curve, ablation studies, statistical significance, inter-annotator agreement, and user testing.

13
How Well Does It Work?
– Evaluation

In this chapter, we will address the question of quantifying how well a **natural language understanding** (**NLU**) system works. Throughout this book, we assumed that we want the NLU systems that we develop to do a good job on the tasks that they are designed for. However, we haven't dealt in detail with the tools that enable us to tell how well a system works – that is, how to evaluate it. This chapter will illustrate a number of evaluation techniques that will enable you to tell how well the system works, as well as to compare systems in terms of performance. We will also look at some ways to avoid drawing erroneous conclusions from evaluation metrics.

The topics we will cover in this chapter are as follows:

- Why evaluate an NLU system?
- Evaluation paradigms
- Data partitioning
- Evaluation metrics
- User testing
- Statistical significance of differences
- Comparing three text classification methods

We will start by asking the question of why it's important to evaluate NLU systems.

Why evaluate an NLU system?

There are many questions that we can ask about the overall quality of an NLU system, and evaluating it is the way that we answer these questions. How we evaluate depends on the goal of developing the system and what we want to learn about the system to make sure that the goal is achieved.

Different kinds of developers will have different goals. For example, consider the goals of the following types of developers:

- I am a researcher, and I want to learn whether my ideas advance the science of NLU. Another way to put this is to ask how my work compares to the **state of the art** (**SOTA**) – that is, the best results that anyone has reported on a particular task.

- I am a developer, and I want to make sure that my overall system performance is good enough for an application.

- I am a developer, and I want to see how much my changes improve a system.

- I am a developer, and I want to make sure my changes have not decreased a system's performance.

- I am a researcher or developer who wants to know how my system performs on different classes of data.

The most important question for all of these developers and researchers is, *how well does the system perform its intended function?*

This is the question we will focus on in this chapter, and we will address how each of these different kinds of developers discovers the information they need. However, there are other important NLU system properties that can be evaluated, and sometimes, these will be more important than overall system performance. It is worth mentioning them briefly so that you are aware of them. For example, other aspects of evaluation questions include the following:

- **The size of the machine learning model that supports the application**: Today's models can be very large, and there is significant research effort directed at making models smaller without significantly degrading their performance. If it is important to have a small model, you will want to look at the trade-offs between model size and accuracy.

- **Training time**: Some algorithms require training time on the order of several weeks on highly capable GPU processors, especially if they are training on large datasets. Reducing this time makes it much easier to experiment with alternative algorithms and tuning hyperparameters. In theory, larger models will provide better results, at the cost of more training time, but in practice, we need to ask how much difference do they make in the performance of any particular task?

- **Amount of training data**: Today's **large language models** (**LLMs**) require enormous amounts of training data. In fact, in the current SOTA, this amount of data is prohibitively large for all but the largest organizations. However, as we saw in *Chapter 11*, LLMs can be fine-tuned with application-specific data. The other consideration for training data is whether there is enough data available to train a system that works well.

- **The expertise of developers**: Depending on highly expert developers is expensive, so a development process that can be performed by less expert developers is usually desirable. The rule-based systems discussed in *Chapter 8*, often require highly expert developers, which is one reason that they tend to be avoided if possible. On the other hand, experimenting with

SOTA deep learning models can call for the knowledge of expert data scientists, who can also be expensive and hard to find.

- **Cost of training**: The training costs for very large models are in the range of millions of dollars, even if we are only taking into account the cost of computing resources. A lower training cost is clearly a desirable property of an NLU system.

- **Environmental impact**: Closely related to the cost of training is its environmental impact in terms of energy expenditure, which can be very high. Reducing this is obviously desirable.

- **Processing time for inference**: This question relates to how long it takes a trained system to process an input and deliver results. With today's systems, this is not usually a problem, with the short inputs that are used with interactive systems such as chatbots or spoken dialog systems. Almost any modern approach will enable them to be processed quickly enough that users will not be annoyed. However, with offline applications such as analytics, where an application may need to extract information from many hours of audio or gigabytes of text, slow processing times will add up.

- **Budget**: Paid cloud-based LLMs such as GPT-4 usually provide very good results, but a local open source model such as BERT could be much cheaper and give results that are good enough for a specific application.

Even though these properties can be important to know when we're deciding on which NLU approach to use, how well an NLU system works is probably the most important. As developers, we need the answers to such fundamental questions as the following:

- *Does this system perform its intended functions well enough to be useful?*

- *As changes are made in the system, is it getting better?*

- *How does this system's performance compare to the performance of other systems?*

The answers to these questions require evaluation methods that assign numerical values to a system's performance. Subjective or non-quantitative evaluation, where a few people look at a system's performance and decide whether it *looks good* or not, is not precise enough to provide reliable answers to those questions.

We will start our discussion of evaluation by reviewing some overall approaches to evaluation, or *evaluation paradigms*.

Evaluation paradigms

In this section, we will review some of the major evaluation paradigms that are used to quantify system performance and compare systems.

Comparing system results on standard metrics

This is the most common evaluation paradigm and probably the easiest to carry out. The system is simply given data to process, and its performance is evaluated quantitatively based on standard metrics. The upcoming *Evaluation metrics* section will delve into this topic in much greater detail.

Evaluating language output

Some NLU applications produce natural language output. These include applications such as translation or summarizing text. They differ from applications with a specific right or wrong answer, such as classification and slot filling, because there is no single correct answer – there could be many good answers.

One way to evaluate machine translation quality is for humans to look at the original text and the translation and judge how accurate it is, but this is usually too expensive to be used extensively. For that reason, metrics have been developed that can be applied automatically, although they are not as satisfactory as human evaluation. We will not cover these in detail here, but we will briefly list them so that you can investigate them if you need to evaluate language output.

The **bilingual evaluation understudy** (BLEU) metric is one well-established metric to evaluate translations. This metric is based on comparing machine translation results to human translations and measuring the difference. Because it is possible that a very good machine translation will be quite different from any particular human translation, the BLEU scores will not necessarily correspond to human judgments of translation quality. Other evaluation metrics for applications that produce language output include the **metric for evaluation for translation with explicit ordering** (METEOR), the **recall-oriented understudy for gisting evaluation** (ROUGE), and the **cross-lingual optimized metric for evaluation of translation** (COMET).

In the next section, we will discuss an approach to evaluation that involves removing part of a system to determine what effect it has on the results. Does it make the results better or worse, or does it not result in any changes?

Leaving out part of a system – ablation

If an experiment includes a pipeline of several operations, it's often informative to compare results by removing steps in the pipeline. This is called **ablation**, and it is useful in two different situations.

The first case is when an experiment is being done for a research paper or an academic project that includes some innovative techniques in the pipeline. In that case, you want to be able to quantify what effect every step in the pipeline had on the final outcome. This will allow readers of the paper to evaluate the importance of each step, especially if the paper is attempting to show that one or more of the steps is a major innovation. If the system still performs well when the innovations are removed, then they are unlikely to be making a significant contribution to the system's overall performance. Ablation studies will enable you to find out exactly what contributions are made by each step.

The second case for ablation is more practical and occurs when you're working on a system that needs to be computationally efficient for deployment. By comparing versions of a system with and without specific steps in the pipeline, you can decide whether the amount of time that they take justifies the degree of improvement they make in the system's performance.

An example of why you would want to consider doing an ablation study for this second reason could include finding out whether preprocessing steps such as stopword removal, lemmatization, or stemming make a difference in the system's performance.

Another approach to evaluation involves a test where several independent systems process the same data and the results are compared. These are shared tasks.

Shared tasks

The field of NLU has long benefited from comparing systems on **shared tasks**, where systems developed by different developers are all tested on a single set of shared data on a specific topic and the results are compared. In addition, the teams working on the shared task usually publish system descriptions that provide very useful insights into how their systems achieved their results.

The shared task paradigm has two benefits:

- First, it enables developers who participate in the shared task to get precise information about how their system compares to others because the data is exactly the same

- Second, the data used in the shared tasks is made available to the research community for use in developing future systems

Shared data can be useful for a long time because it is not affected by changes in NLU technology – for example, the **air travel information system** (**ATIS**) data from a travel planning task has been in use since the early 1990s.

The *NLP-progress* website at `http://nlpprogress.com/` is a good source of information on shared tasks and shared task data.

In the next section, we will look at ways of dividing data into subsets in preparation for evaluation.

Data partitioning

In earlier chapters, we divided our datasets into subsets used for *training*, *validation*, and *testing*.

As a reminder, training data is used to develop the NLU model that is used to perform the eventual task of the NLU application, whether that is classification, slot-filling, intent recognition, or most other NLU tasks.

Validation data (sometimes called **development test** data) is used during training to assess the model on data that was not used in training. This is important because if the system is tested on the training data, it could get a good result simply by, in effect, memorizing the training data. This would be misleading because that kind of system isn't very useful – we want the system to generalize or work well on the new data that it's going to get when it is deployed. Validation data can also be used to help tune hyperparameters in machine learning applications, but this means that during development, the system has been exposed a bit to the validation data, and as a consequence, that data is not as novel as we would like it to be.

For that reason, another set of completely new data is typically held out for a final test; this is the test data. In preparation for system development, the full dataset is partitioned into training, validation, and test data. Typically, around 80% of the full dataset is allocated to training, 10% is allocated to validation, and 10% is allocated to testing.

There are three general ways that data can be partitioned:

- In some cases, the dataset is already partitioned into training, validation, and testing sets. This is most common in generally available datasets such as those we have used in this book or shared tasks. Sometimes, the data is only partitioned into training and testing. If so, a subset of the training data should be used for validation. Keras provides a useful utility, `text_dataset_from_directory`, that loads a dataset from a directory, using subdirectory names for the supervised categories of the text in the directories, and partitions a validation subset. In the following code, the training data is loaded from the `aclImdb/train` directory, and then 20% of the data is split out to be used as validation data:

```
raw_train_ds = tf.keras.utils.text_dataset_from_directory(
    'aclImdb/train',
    batch_size=batch_size,
    validation_split=0.2,
    subset='training',
    seed=seed)
```

- Of course, making use of a previous partition is only useful when you work with generally available datasets that have already been partitioned. If you have your own data, you will need to do your own partitioning. Some of the libraries that we've been working with have functions that can automatically partition the data when it is loaded. For example, scikit-learn, TensorFlow, and Keras have `train_test_split` functions that can be used for this.

- You can also manually write Python code to partition your dataset. Normally, it would be preferable to work with pretested code from a library, so writing your own Python code for this would not be recommended unless you are unable to find a suitable library, or you are simply interested in the exercise of learning more about the partitioning process.

The final data partitioning strategy is called **k-fold cross-validation**. This strategy involves partitioning the whole dataset into k subsets, or *folds*, and then treating each of the folds as test data for an evaluation of the system. The overall system score in k-fold cross-validation is the average score of all of the tests.

The advantage of this approach is that it reduces the chances of an accidental difference between the test and training data, leading to a model that gives poor predictions for new test examples. It is harder for an accidental difference to affect the results in k-fold cross-validation because there is no strict line between training and test data; the data in each fold takes a turn at being the test data. The disadvantage of this approach is that it multiplies the amount of time it takes to test the system by k, which could become very large if the dataset is also large.

Data partitioning is necessary for almost every evaluation metric that we will discuss in this chapter, with the exception of user testing.

In the next section, we will look at some of the most common specific quantitative metrics. In *Chapter 9*, we went over the most basic and intuitive metric, accuracy. Here, we will review other metrics that can usually provide better insights than accuracy and explain how to use them.

Evaluation metrics

There are two important concepts that we should keep in mind when selecting an evaluation metric for NLP systems or, more generally, any system that we want to evaluate:

- **Validity**: The first is validity, which means that the metric corresponds to what we think of intuitively as the actual property we want to know about. For example, we wouldn't want to pick the length of a text as a measurement for its positive or negative sentiment because the length of a text would not be a valid measure of its sentiment.

- **Reliability**: The other important concept is reliability, which means that if we measure the same thing repeatedly, we always get the same result.

In the next sections, we will look at some of the most commonly used metrics in NLU that are considered to be both valid and reliable.

Accuracy and error rate

In *Chapter 9*, we defined accuracy as the number of correct system responses divided by the overall number of inputs. Similarly, we defined **error rate** as incorrect responses divided by the number of inputs. Note that you might encounter a **word error rate** when you read reports of speech recognition results. Because the word error rate is calculated according to a different formula that takes into account different kinds of common speech recognition errors, this is a different metric, which we will not cover here.

The next section will go over some more detailed metrics, precision, recall, and F_1 score, which were briefly mentioned in *Chapter 9*.

These metrics usually provide more insight into a system's processing results than accuracy.

Precision, recall, and F₁

Accuracy can give misleading results under certain conditions. For example, we might have a dataset where there are 100 items but the classes are unbalanced, and the vast majority of cases (say, 90) belong to one class, which we call the majority class. This is very common in real datasets. In that situation, 90% accuracy could be obtained by automatically assigning every case to the majority class, but the 10 remaining instances that belong to the other class (let's assume we only have two classes for simplicity) would always be wrong. Intuitively, accuracy would seem to be an invalid and misleading metric for this situation. To provide a more valid metric, some refinements were introduced – most importantly, the concepts of recall and precision.

Recall means that a system finds every example of a class and does not miss any. In our example, it correctly finds all 90 instances of the majority class but no instances of the other class.

The formula for recall is shown in the following equation, where *true positives* are instances of the class that were correctly identified, and *false negatives* are instances of that class that were missed. In our initial example, assuming that there are 100 items in the dataset, the system correctly found 90 (the majority class) but missed the 10 examples of the other class. Therefore, the recall score of the majority class is 1.0, but the recall score of the other class is 0:

$$recall = \frac{true\ positives}{true\ positives + false\ negatives}$$

On the other hand, **precision** means that whatever was identified was correct. Perfect precision means that no items were misidentified, although some items might have been missed. The following is the formula for precision, where *false positives* are instances that were misidentified. In this example, the false positives are the 10 items that were incorrectly identified as belonging to the majority class. In our example, the precision score for the majority class is 1.0, but for the other class, it is 0 because there are no *true positives*. The precision and recall scores in this example give us more detail about the kinds of mistakes that the system has made:

$$precision = \frac{true\ positives}{true\ positives + false\ positives}$$

Finally, there is another important metric, **F₁**, that combines precision and recall scores. This is a useful metric because we often want to have a single overall metric to describe a system's performance. This is probably the most commonly used metric in NLU. The formula for F1 is as follows:

$$F_1 = \frac{precision \times recall}{precision + recall}$$

Figure 13.1 shows correct and incorrect classifications graphically for one class. Items in the ellipse are identified as **Class 1**. The round markers inside the ellipse are true positives – items identified as **Class 1** that are actually **Class 1**. The square markers inside the ellipse are false positives – items in **Class 2** that are incorrectly identified as **Class 1**. Round markers outside of the ellipse are false negatives – items that are actually in **Class 1** but were not recognized:

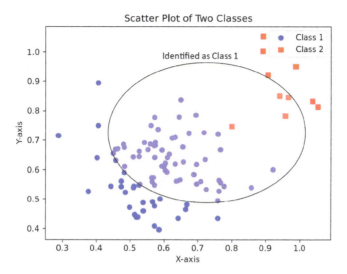

Figure 13.1 – The classification for one class

There is an assumption with the F_1 metric that the recall and precision scores are equally important. This is true in many cases, but not always. For example, consider an application whose goal is to detect mentions of its company's products in tweets. The developers of this system might consider it to be very important not to miss any tweets about their products. In that case, they will want to emphasize recall at the expense of precision, which means that they might get a lot of false positives – tweets that the system categorized as being about their product but actually were not. There are more general versions of the F_1 metric that can be used to weight recall and precision if you develop a system where recall and precision are not equally important.

The receiver operating characteristic and area under the curve

The decision of the class to which a test item belongs depends on its score for that class. If the item's score exceeds a given threshold score (selected by the developer), then the system decides that the item does actually belong in that class. We can see that true positive and false positive scores can be affected by this threshold. If we make the threshold very high, few items will fall into that class, and the true positives will be lower. On the other hand, if the threshold is very low, the system will decide that many items fall into that class, the system will accept many items that don't belong, and the false positives will be very high. What we really want to know is how good the system is at discriminating between classes at every threshold.

A good way to visualize the trade-off between false positives (precision failures) and false negatives (recall failures) is a graph called the **receiver operating characteristic** (**ROC**) and its related metric, the **area under the curve** (**AUC**). The ROC curve is a measurement of how good a system is overall at discriminating between classes. The best way to understand this is to look at an example of an ROC curve, which we can see in *Figure 13.2*. This figure is based on randomly generated data:

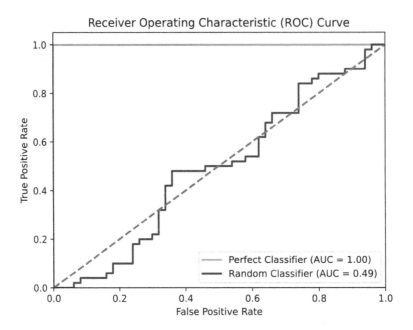

Figure 13.2 – The ROC curve for a perfect classifier compared to a random classifier

In *Figure 13.2*, we can see that at point (**0**, **1**), the system has no false positives, and all of the true positives are correctly detected. If we set the acceptance threshold to **1**, the system will still accept all of the true positives, and there will still be no false positives. On the other hand, if the classifier can't tell the classes apart at all, we would get something like the **Random Classifier** line. No matter where we set the threshold, the system will still make many mistakes, as if it was randomly assigning classes to inputs.

A common way to summarize a system's ability to discriminate between classes is the area under the ROC curve, or AUC. The perfect classifier's AUC score will be **1.0**, a random classifier's AUC score will be around **0.5**, and if the classifier performs at a level that is worse than random, the AUC score will be less than **0.5**. Note that we look at a binary classification problem here because it is simpler to explain, but there are techniques to apply these ideas to multi-class problems as well. We will not go into these here, but a good discussion of the techniques to look at ROC curves for multi-class datasets can be found in the scikit-learn documentation at `https://scikit-learn.org/stable/auto_examples/model_selection/plot_roc.html`.

Confusion matrix

Another important evaluation tool is a **confusion matrix**, which shows how often each class is confused with each other class. This will be much clearer with an example, so we will postpone discussing confusion matrices until we go over the example at the end of this chapter.

User testing

In addition to direct system measurements, it is also possible to evaluate systems with user testing, where test users who are representative of a system's intended users interact with it.

User testing is a time-consuming and expensive type of testing, but sometimes, it is the only way that you can find out qualitative aspects of system performance – for example, how easy it is for users to complete tasks with a system, or how much they enjoy using it. Clearly, user testing can only be done on aspects of the system that users can perceive, such as conversations, and users should be only expected to evaluate the system as a whole – that is, users can't be expected to reliably discriminate between the performance of the speech recognition and the NLU components of the system.

Carrying out a valid and reliable evaluation with users is actually a psychological experiment. This is a complex topic, and it's easy to make mistakes that make it impossible to draw conclusions from the results. For those reasons, providing complete instructions to conduct user testing is outside of the scope of this book. However, you can do some exploratory user testing by having a few users interact with a system and measuring their experiences. Some simple measurements that you can collect in user testing include the following:

- Having users fill out a simple questionnaire about their experiences.

- Measuring how long users spend interacting with the system. The amount of time users spend interacting with the system can be a positive or negative measurement, depending on the purpose of the system, such as the following:

  - If you are developing a social system that's just supposed to be fun, you will want them to spend more time interacting with the system

  - On the other hand, if you are developing a task-oriented system that will help users complete a task, you will usually want the users to spend less time interacting with the system

However, you also have to be cautious with user testing:

- It is important for users to be representative of the actual intended user population. If they are not, the results can be very misleading. For example, a company chatbot intended for customer support should be tested on customers, not employees, since employees have much more knowledge about the company than customers and will ask different questions.

- Keep questionnaires simple, and don't ask users to provide information that isn't relevant to what you are trying to learn. Users who are bored or impatient with the questionnaire will not provide useful information.

- If the results of user testing are very important to the project, you should find a human factors engineer – that is, someone with experience designing experiments with human users – to design the testing process so that the results are valid and reliable.

So far, we have looked at several different ways to measure the performance of systems. Applying each of these techniques will result in one or more numerical values, or *metrics* that quantify a system's performance. Sometimes, when we use these metrics to compare several systems, or several versions of the same system, we will find that the different values of the metrics are small. It is then worth asking whether small differences are really meaningful. This question is addressed by the topic of statistical significance, which we will cover in the next section.

Statistical significance of differences

The last general topic we will cover in evaluation is the topic of determining whether the differences between the results of experiments we have done reflect a real difference between the experimental conditions, or whether they reflect differences that are due to chance. This is called **statistical significance**. Whether a difference in the values of the metrics represents a real difference between systems isn't something that we can know for certain, but what we can know is how likely it is that a difference that we're interested in is due to chance. Let's suppose we have the situation with our data that's shown in *Figure 13.3*:

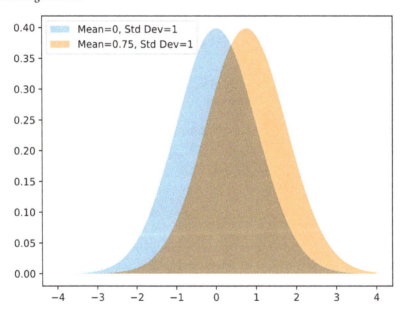

Figure 13.3 – Two distributions of measurement values – do they reflect
a real difference between the things they're measuring?

Figure 13.3 shows two sets of measurements, one with a mean of **0**, on the left, and one with a mean of **0.75**, on the right. Here, we might be comparing the performance of two classification algorithms on different datasets, for example. It looks like there is a real difference between the algorithms, but could this difference be an accident? We usually consider that if the probability of a difference of this size occurring by chance is once out of 20, then the difference is considered to be statistically significant, or not due to chance. Of course, that means that 1 out of 20 statistically significant results is probably actually due to chance. This probability is determined by standard statistical formulas such as the *t-statistic* or the *analysis of variance*.

If you read a paper that performs a significance test on its results and it states something such as *p<.05*, the *p* refers to the probability of the difference being due to chance. This kind of statistical analysis is most commonly performed with data where knowing whether differences are statistically significant is very important, such as academic papers for presentation at conferences or for publication in academic journals. We will not cover the process of how statistical significance is calculated here, since it can become quite a complex process, but you should be aware of what it means if you have a need for it.

It is also important to consider that a result can be statistically significant but not really meaningful in realistic situations. If the results of one classification algorithm are only slightly better than another but the better algorithm is much more complicated to compute, it might not be worth bothering with the better algorithm. These trade-offs are something that has to be considered from the perspective of how the algorithm will be used. Even a small, but significant, difference could be important in an academic paper but not important at all in a deployed application.

Now that we have reviewed several approaches to evaluating systems, let's take a look at applying them in practice. In the next section, we will work through a case study that compares three different approaches to text classification on the same data.

Comparing three text classification methods

One of the most useful things we can do with evaluation techniques is to decide which of several approaches to use in an application. Are the traditional approaches such as **term frequency - inverse document frequency (TF-IDF)**, **support vector machines (SVMs)**, and **conditional random fields (CRFs)** good enough for our task, or will it be necessary to use deep learning and transformer approaches that have better results at the cost of longer training time?

In this section, we will compare the performance of three approaches on a larger version of the movie review dataset that we looked at in *Chapter 9*. We will look at using a small BERT model, TF-IDF vectorization with the Naïve Bayes classification, and a larger BERT model.

A small transformer system

We will start by looking at the BERT system that we developed in *Chapter 11*. We will use the same BERT model as in *Chapter 11*, which is one of the smallest BERT models, `small_bert/bert_en_uncased_L-2_H-128_A-2`, with two layers, a hidden size of `128`, and two attention heads.

We will be making a few changes in order to better evaluate this model's performance.

First, we will add new metrics, precision and recall, to the `BinaryAccuracy` metric that we used in *Chapter 11*:

```
metrics = [tf.metrics.Precision(),tf.metrics.Recall(),tf.metrics.
BinaryAccuracy()]
```

With these metrics, the `history` object will include the changes in precision and recall during the 10-epoch training process. We can look at these with the following code:

```
history_dict = history.history
print(history_dict.keys())

acc = history_dict['binary_accuracy']
precision = history_dict['precision_1']
val_acc = history_dict['val_binary_accuracy']
loss = history_dict['loss']
val_recall = history_dict['val_recall_1']
recall = history_dict['recall_1']
val_loss = history_dict['val_loss']
val_precision = history_dict['val_precision_1']
precision = history_dict['precision_1']
val_recall = history_dict['val_recall_1']
```

As shown in the preceding code snippet, the first steps are to pull out the results we are interested in from the `history` object. The next step is to plot the results as they change over training epochs, which is calculated in the following code:

```
epochs = range(1, len(acc) + 1)
fig = plt.figure(figsize=(10, 6))
fig.tight_layout()

plt.subplot(4, 1, 1)
# r is for "solid red line"
plt.plot(epochs, loss, 'r', label='Training loss')
# b is for "solid blue line"
plt.plot(epochs, val_loss, 'b', label='Validation loss')
plt.title('Training and validation loss')
# plt.xlabel('Epochs')
plt.ylabel('Loss')
plt.legend()
```

The preceding code shows the plotting process for the training and validation loss, which results in the top plot in *Figure 13.4*. The rest of the plots in *Figure 13.4* and the plots in *Figure 13.5* are calculated in the same manner, but we will not show the full code here, since it is nearly the same as the code for the first plot.

Figure 13.4 shows decreasing loss and increasing accuracy over training epochs, which we saw previously. We can see that loss for both the training and validation data is leveling off between **0.40** and **0.45**. At epoch **7**, it looks like additional training is unlikely to improve performance:

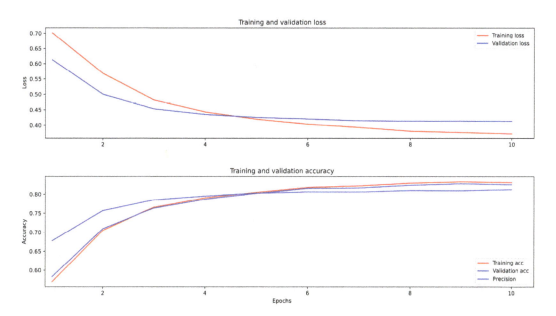

Figure 13.4 – Loss and accuracy over 10 training epochs

Since we have added the precision and recall metrics, we can also see that these metrics level off at around **0.8** at around **7** epochs of training, as shown in *Figure 13.5*:

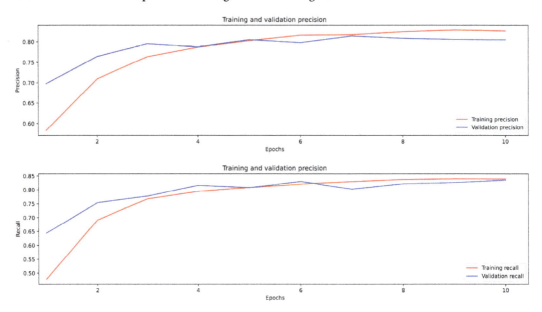

Figure 13.5 – Precision and recall over 10 training epochs

We can also look at the confusion matrix and classification report for more information with the following code, using functions from scikit-learn:

```
# Displaying the confusion matrix
%matplotlib inline
from sklearn.metrics import confusion_
matrix,ConfusionMatrixDisplay,f1_score,classification_report
import matplotlib.pyplot as plt
plt.rcParams.update({'font.size': 12})

disp = ConfusionMatrixDisplay(confusion_matrix = conf_matrix,
                                          display_labels = class_
names)
print(class_names)
disp.plot(xticks_rotation=75,cmap=plt.cm.Blues)

plt.show()
```

This will plot the confusion matrix as shown in *Figure 13.6*:

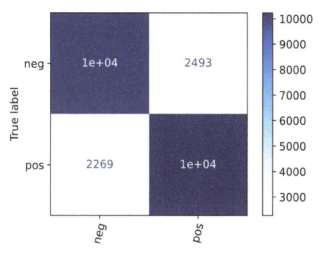

Figure 13.6 – The confusion matrix for a small BERT model

The dark cells in *Figure 13.6* show the number of correct classifications, where actual negative and positive reviews were assigned to the right classes. We can see that there are quite a few incorrect classifications. We can see more detail by printing a summary classification report with the following code:

```
print(classification_report(y_test, y_pred, target_names=class_names))
['neg', 'pos']
```

The classification report shows the `recall`, `precision`, and `F1` scores for both classes:

	precision	recall	f1-score	support
neg	0.82	0.80	0.81	12501
pos	0.80	0.82	0.81	12500
accuracy			0.81	25001
macro avg	0.81	0.81	0.81	25001
weighted avg	0.81	0.81	0.81	25001

From this report, we can see that the system is almost equally good at recognizing positive and negative reviews. The system correctly classifies many reviews but is still making many errors.

Now, we will compare the BERT test with one of our earlier tests, based on TF-IDF vectorization and Naïve Bayes classification.

TF-IDF evaluation

In *Chapter 9*, we learned about vectorizing with TF-IDF and classifying with Naïve Bayes. We illustrated the process with the movie review corpus. We want to compare these very traditional techniques with the newer transformer-based LLMs, such as BERT, that we just saw. How much better are transformers than the traditional approaches? Do the transformers justify their much bigger size and longer training time? To make the comparison between BERT and TF-IDF/Naïve Bayes fair, we will use the larger `aclimdb` movie review dataset that we used in *Chapter 11*.

We will use the same code we used in *Chapter 9* to set up the system and perform the TF-IDF/Naïve Bayes classification, so we won't repeat it here. We will just add the final code to show the confusion matrix and display the results graphically:

```
# View the results as a confusion matrix
from sklearn.metrics import confusion_matrix
conf_matrix = confusion_matrix(labels_test, labels_
pred,normalize=None)
print(conf_matrix)
[[9330 3171]
 [3444 9056]]
```

The text version of the confusion matrix is the array:

```
[[9330 3171]

[3444 9056]]
```

However, this is not very easy to understand. We can display it more clearly by using `matplotlib` in the following code:

```
# Displaying the confusion matrix
from sklearn.metrics import confusion_
matrix,ConfusionMatrixDisplay,f1_score,classification_report
import matplotlib.pyplot as plt
plt.rcParams.update({'font.size': 12})

disp = ConfusionMatrixDisplay(confusion_matrix = conf_matrix, display_
labels = class_names)
print(class_names)
disp.plot(xticks_rotation=75,cmap=plt.cm.Blues)

plt.show()
```

The result is shown in *Figure 13.7*:

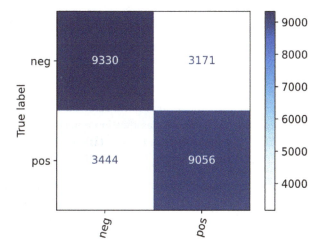

Figure 13.7 – The confusion matrix for TF-IDF/the Naïve Bayes classification

The dark cells in *Figure 13.7* show the correct classifications, where actual negative and positive reviews were assigned to the right classes. We can also see that **3171** actual negative reviews were misclassified as positive and **3444** actual positive reviews were misclassified as negative.

To see the recall, precision, and F_1 scores, we can print the classification report:

```
print(classification_report(labels_test, labels_pred, target_
names=class_names))
```

The resulting classification report shows the recall, precision, and F_1 scores, along with the accuracy, the number of items in each class (`support`), and other statistics:

	precision	recall	f1-score	support
neg	0.73	0.75	0.74	12501
pos	0.74	0.72	0.73	12500
accuracy			0.74	25001
macro avg	0.74	0.74	0.74	25001
weighted avg	0.74	0.74	0.74	25001

From this report, we can see that the system is slightly better at recognizing negative reviews. The system correctly classifies many reviews but is still making many errors. Comparing these results to the results for the BERT system in *Figure 13.6*, we can see that the BERT results are quite a bit better. The BERT F_1 score is **0.81**, while the TF-IDF/Naïve Bayes F_1 score is **0.74**. For this task, the BERT system would be the better choice, but can we do better?

In the next section, we will ask another question that compares two systems. The question concerns what might happen with other BERT transformer models. Larger models will almost always have better performance than smaller models but will take more training time. How much better will they be in terms of performance?

A larger BERT model

So far, we have compared a very small BERT model to a system based on TF-IDF and Naïve Bayes. We saw from the comparison between the two systems' classification reports that the BERT system was definitely better than the TF-IDF/Naïve Bayes system. For this task, BERT is better as far as correct classification goes. On the other hand, BERT is much slower to train.

Can we get even better performance with another transformer system, such as another variation of BERT? We can find this out by comparing our system's results to the results from other variations of BERT. This is easy to do because the BERT system includes many more models of various sizes and complexities, which can be found at `https://tfhub.dev/google/collections/bert/1`.

Let's compare another model to the BERT system we just tested. The system that we've tested is one of the smaller BERT systems, `small_bert/bert_en_uncased_L-2_H-128_A-2`. The name encodes its important properties – two layers, a hidden size of `128`, and two attention heads. We might be interested in finding out what happens with a larger model. Let's try one, `small_bert/bert_en_uncased_L-4_H-512_A-8`, which is nevertheless not too large to train on a CPU (rather than a GPU). This model is still rather small, with four layers, a hidden size of `512`, and eight attention heads. Trying a different model is quite easy to do, with just a minor modification of the code that sets up the BERT model for use by including the information for the new model:

```
bert_model_name = 'small_bert/bert_en_uncased_L-4_H-512_A-8'

map_name_to_handle = {
    'small_bert/bert_en_uncased_L-4_H-512_A-8' :
        'https://tfhub.dev/tensorflow/small_bert/bert_en_uncased_L-
4_H-512_A-8/1',
}
map_model_to_preprocess = {
    'small_bert/bert_en_uncased_L-4_H-512_A-8':
        'https://tfhub.dev/tensorflow/bert_en_uncased_
preprocess/3',
}
```

As the code shows, all we have to do is select the model's name, map it to the URL where it is located, and assign it to a preprocessor. The rest of the code will be the same. *Figure 13.8* shows the confusion matrix for the larger BERT model:

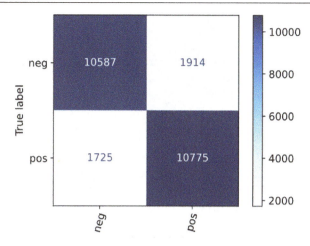

Figure 13.8 – The confusion matrix for the larger BERT model

The performance of this model, as shown in the confusion matrix, is superior to that of the smaller BERT model as well as the TF-IDF/Naïve Bayes model. The classification report is shown in the following snippet, and it also shows that this model performs the best of the three models we have looked at in this section, with an average F1 score of 0.85, compared to 0.81 for the smaller BERT model and 0.74 for the TF-IDF/Naïve Bayes model.

```
                precision    recall     f1-score     support

    neg            0.86        0.85        0.85        12501
    pos            0.85        0.86        0.86        12500

    accuracy                               0.85        25001
    macro avg      0.85        0.85        0.85        25001
    weighted avg   0.85        0.85        0.85        25001
```

The training time for this model on the `aclimdb` dataset with 25,000 items was about eight hours on a standard CPU, which is probably acceptable for most applications. Is this performance good enough? Should we explore even bigger models? There is clearly quite a bit of room for improvement, and there are many other BERT models that we can experiment with. The decision of whether the results are acceptable or not is up to the developers of the application and depends on the importance of getting correct answers and avoiding wrong answers. It can be different for every application. You are encouraged to experiment with some of the larger models and consider whether the improved performance justifies the additional training time.

Summary

In this chapter, you learned about a number of important topics related to evaluating NLU systems. You learned how to separate data into different subsets for training and testing, and you learned about the most commonly used NLU performance metrics – accuracy, precision, recall, F_1, AUC, and confusion matrices – and how to use these metrics to compare systems. You also learned about related topics, such as comparing systems with ablation, evaluation with shared tasks, statistical significance testing, and user testing.

The next chapter will start *Part 3* of this book, where we cover systems in action – applying NLU at scale. We will start *Part 3* by looking at what to do if a system isn't working. If the original model isn't adequate or the system models a real-world situation that changes, what has to be changed? The chapter discusses topics such as adding new data and changing the structure of the application.

Part 3:
Systems in Action – Applying Natural Language Understanding at Scale

In *Part 3*, you will learn about applying natural language understanding in running applications. This part will cover adding new data to existing applications, dealing with volatile applications, adding and removing categories, and will include a final chapter summarizing the book and looking to the future of natural language understanding.

We focus on getting NLU systems out of the lab and making them do real work solving practical problems.

This part comprises the following chapters:

- *Chapter 14, What to Do If the System Isn't Working*
- *Chapter 15, Summary and Looking to the Future*

14

What to Do If the
System Isn't Working

In this chapter, we will discuss how to improve systems. If the original model's first round of training fails to produce a satisfactory performance or the real-world scenario that the system addresses undergoes changes, we need to modify something to enhance the system's performance. In this chapter, we will discuss techniques such as adding new data and changing the structure of an application, while at the same time ensuring that new data doesn't degrade the performance of the existing system. Clearly, this is a big topic, and there is a lot of room to explore how to improve the performance of **natural language understanding** (NLU) systems. It isn't possible to cover all the possibilities here, but this chapter should give you a good perspective on the most important options and techniques that can improve system performance.

We will cover the following topics in this chapter:

- Figuring out that a system isn't working

- Fixing accuracy problems

- Moving on to deployment

- Problems after deployment

The first step is to find out that a system isn't working as well as desired. This chapter will include a number of examples of tools that can help with this. We will start by listing the software requirements needed to run these examples.

Technical requirements

We will be using the following data and software to run the examples in this chapter:

- Our usual development environment – that is, Python 3 and Jupyter Notebook

- The TREC dataset

- The Matplotlib and Seaborn packages, which we will use to display graphical charts

- pandas and NumPy for numerical manipulation of data

- The BERT NLU system, previously used in *Chapter 11* and *Chapter 13*

- The Keras machine learning library, for working with BERT

- NLTK, which we will use for generating new data

- An OpenAI API key which we will use to access the OpenAI tools

Figuring out that a system isn't working

Figuring out whether a system isn't working as well as it should be is important, both during initial development as well as during ongoing deployment. We'll start by looking at poor performance during initial development.

Initial development

The primary techniques we will use to determine that our system isn't working as well as we'd like are the evaluation techniques we learned about in *Chapter 13*. We will apply those in this chapter. We will also use confusion matrices to detect specific classes that don't work as well as the other classes.

It is always a good idea to look at the dataset at the outset and check the balance of categories because unbalanced data is a common source of problems. Unbalanced data does not necessarily mean that there will be accuracy problems, but it's valuable to understand our class balance at the beginning. That way, we will be prepared to address accuracy issues caused by class imbalance as system development progresses.

Checking category balance

For our data exploration in this chapter, we will use the **Text Retrieval Conference** (**TREC**) dataset, which is a commonly used multi-class classification dataset and can be downloaded from Hugging Face (`https://huggingface.co/datasets/trec`).

Dataset citations

Learning Question Classifiers, Li, Xin and Roth, Dan, *{COLING} 2002: The 19th International Conference on Computational Linguistics*, 2002, `https://www.aclweb.org/anthology/C02-1150`

Toward Semantics-Based Answer Pinpointing, Hovy, Eduard and Gerber, Laurie and Hermjakob, Ulf and Lin, Chin-Yew and Ravichandran, Deepak, *Proceedings of the First International Conference on Human Language Technology Research*, 2001, `https://www.aclweb.org/anthology/H01-1069`

The dataset consists of 5,452 training examples of questions that users might ask of a system and 500 test examples. The goal of the classification task is to identify the general topic of a question as the first step in answering it. The question topics are organized into two levels, consisting of six broad categories and 50 more specific subcategories that fall under the broader topics.

We will be working with the broad categories, which are as follows:

- Abbreviation (ABBR)

- Description (DESC)

- Entity (ENTY)

- Human (HUM)

- Location (LOC)

- Number (NUM)

One important task at the beginning is to find out how many documents are in each class. We want to see whether all of the classes have enough texts for effective training and whether no classes are significantly more or less common than the others.

So far in this book, we have seen many ways to load datasets. One of the easiest ways to load a dataset is based on data being organized into folders, with separate folders for each class. Then, we can load the dataset with the `tf.keras.utils.text_dataset_from_directory()` function, which we used several times in previous chapters, and see the class names. This is shown in the following code:

```
# find out the total number of text files in the dataset and what the
classes are
import tensorflow as tf
import matplotlib.pyplot as plt
import pandas as pd
import seaborn as sns
import numpy as np

training_ds = tf.keras.utils.text_dataset_from_directory(
    'trec_processed/training')
```

```
class_names = training_ds.class_names
print(class_names)
Found 5452 files belonging to 6 classes.
['ABBR', 'DESC', 'ENTY', 'HUM', 'LOC', 'NUM']
```

We can then count the number of files in each class and display them in a bar graph with this code, using the `matplotlib` and `seaborn` graphics libraries:

```
files_dict = {}
for class_name in class_names:
    files_count = training_ds.list_files(
        'trec_processed/training/' + class_name + '/*.txt')
    files_length = files_count.cardinality().numpy()
    category_count = {class_name:files_length}
    files_dict.update(category_count)

# Sort the categories, largest first
from collections import OrderedDict
sorted_files_dict = sorted(files_dict.items(),
    key=lambda t: t[1], reverse=True)
print(sorted_files_dict)

# Conversion to Pandas series
pd_files_dict = pd.Series(dict(sorted_files_dict))

# Setting figure, ax into variables
fig, ax = plt.subplots(figsize=(20,10))

# plot
all_plot = sns.barplot(x=pd_files_dict.index,
    y = pd_files_dict.values, ax=ax, palette = "Set2")
plt.xticks(rotation = 90)
plt.show()
[('ENTY', 1250), ('HUM', 1223), ('ABBR', 1162),
    ('LOC', 896), ('NUM', 835), ('DESC', 86)]
```

While this code prints out the count of texts in each class as text output, it is also very helpful to see the totals as a bar graph. We can use the graphics libraries to create this graph:

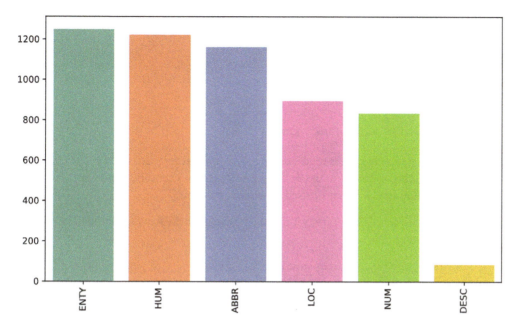

Figure 14.1 – Coarse-grained class counts in the TREC data

As *Figure 14.1* shows, the DESC class is much smaller than the others, and it is possible that there will be accuracy problems with this class. There are ways to address this situation, which is one of the main topics of this chapter, but for now, we won't make any changes until we see that this actually causes a problem.

Doing initial evaluations

Once we have done this initial exploration, we will want to try training one or more initial models for the data and evaluate them using some of the techniques we learned in *Chapter 13*.

For this exploration, we will use the BERT-based training process that was covered in *Chapter 13*, so we won't duplicate that here. However, there are a few changes in the model that we need to make because we are now working with a *categorical* classification problem (six classes), rather than a binary classification problem (two classes), and it is worth pointing these out. We can see the new model definition in the following code:

```
def build_classifier_model():
    text_input = tf.keras.layers.Input(shape=(),
        dtype=tf.string, name='text')
    preprocessing_layer = hub.KerasLayer(
        tfhub_handle_preprocess, name='preprocessing')
    encoder_inputs = preprocessing_layer(text_input)
    encoder = hub.KerasLayer(tfhub_handle_encoder,
```

```
        trainable=True, name='BERT_encoder')
outputs = encoder(encoder_inputs)
net = outputs['pooled_output']
net = tf.keras.layers.Dropout(0.1)(net)
net = tf.keras.layers.Dense(6, activation =
    tf.keras.activations.softmax,
    name='classifier')(net)
return tf.keras.Model(text_input, net)
```

The two changes that are needed in the model definition for the categorical task are in the final layer, which has six outputs, corresponding to the six classes, and a softmax activation function, as opposed to the sigmoid activation function that we used for binary problems.

The other changes that are needed for categorical data are changes in the loss function and the metrics, as shown in the following code:

```
loss="sparse_categorical_crossentropy"
metrics = tf.metrics.CategoricalAccuracy()
```

Here, we will define the categorical loss and metrics functions. Other metrics are available, but we will just look at accuracy here.

After training the model, as we did in *Chapter 13*, we can look at the final scores. If the model does not meet the overall performance expectations for the application using the metrics that have been chosen, you can try different hyperparameter settings, or you can try other models. This was the process we followed in *Chapter 13*, where we compared the performance of three different models on the movie review data.

Keeping in mind that larger models are likely to have better performance, you can try increasing the size of the models. There is a limit to this strategy – at some point, the larger models will become very slow and unwieldy. You might also see that the payoff from larger and larger models becomes smaller, and performance levels off. This probably means that increasing the size of the models will not solve the problem.

There are many different possibilities to look at different hyperparameter settings. This is, in general, a huge search space that can't be fully explored, but there are some heuristics that you can use to find settings that could improve your results. Looking at the training history charts of loss and accuracy changes over epochs should give you a good idea of whether additional training epochs are likely to be helpful. Different batch sizes, learning rates, optimizers, and dropout layers can also be explored.

Another strategy to diagnose system performance is to look at the data itself.

One initial evaluation we can do is a more fine-grained check for weak classes, by looking at the probabilities of the classifications for a large number of items in the dataset. We will look at this in the next section.

Checking for weak classes

Low probabilities for a class of items are a sign that a system is not able to classify items with high confidence and has a good chance of making errors. To check for this, we can use the model to predict the classification of a subset of our data and look at the average scores, as shown in the following code:

```python
import matplotlib.pyplot as plt
import seaborn as sns

scores = [[],[],[],[],[],[]]

for text_batch, label_batch in train_ds.take(100):
    for i in range(160):
        text_to_classify = [text_batch.numpy()[i]]
        prediction = classifier_model.predict(
            text_to_classify)
        classification = np.max(prediction)
        max_index = np.argmax(prediction)
        scores[max_index].append(classification)
averages = []
for i in range(len(scores)):
    print(len(scores[i]))
    averages.append(np.average(scores[i]))
print(averages)
```

This code goes through a subset of the TREC training data, predicts each item's class, saves the predicted class in the `classification` variable, and then adds it to the `scores` list for the predicted class.

The final step in the code is to iterate through the scores list and print the length and average score for each class. The results are shown in *Table 14.1*:

Class	Number of items	Average score
ABBR	792	0.9070532
DESC	39	0.8191106
HUM	794	0.8899161
ENTY	767	0.9638871
LOC	584	0.9767452
NUM	544	0.9651737

Table 14.1 – The number of items and the average score for each class

We can see from *Table 14.1* that the number of items and average probabilities of the class predictions vary quite a bit. As you will recall from the counts we did in *Figure 14.1*, we were already concerned about the DESC class because it was so small relative to the other classes. We can investigate this a bit further by looking at the predicted classifications of the individual items in each class with the following code:

```python
def make_histogram(score_data,class_name):
    sns.histplot(score_data,bins = 100)
    plt.xlabel("probability score")
    plt.title(class_name)
    plt.show()
for i in range(len(scores)):
    make_histogram(scores[i],class_names[i])
```

Let's look at the histograms for the DESC and LOC classes, which are at the extreme ends of the set of average scores. The LOC class is shown in *Figure 14.2*:

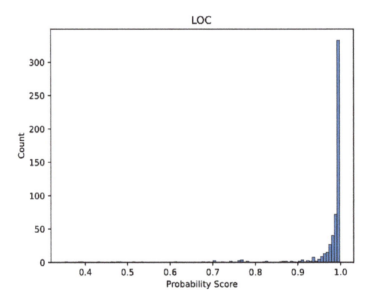

Figure 14.2 – The distribution of probability scores for the LOC class

We can see in *Figure 14.2* that not only is the average probability very high (which we saw in *Table 14.1* as well) but there also are very few probabilities under **0.9** in the LOC class. This class is likely to be very accurate in the deployed application.

There is a second, less obvious advantage to classes that show the pattern in *Figure 14.2*. In a deployed interactive application, we don't want a system to give users answers that it's not very confident of. This is because they're more likely to be wrong, which would mislead the users. For that reason, developers should define a *threshold* probability score, which an answer has to exceed before the system provides that answer to the user.

If the probability is lower than the threshold, the system should respond to the user that it doesn't know the answer. The value of the threshold has to be set by the developer, based on the trade-off between the risk of giving users wrong answers and the risk of annoying users by saying *I don't know* too frequently. *In Figure 14.2* we can see that if we set the threshold to **0.9**, the system will not have to say *I don't know* very often, which will improve user satisfaction with the system.

Let's contrast *Figure 14.2* with a histogram for the DESC class, which we can see in *Figure 14.3*:

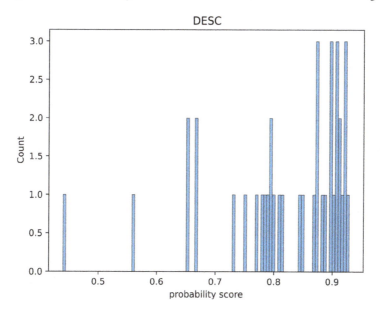

Figure 14.3 – The distribution of probability scores for the DESC class

Figure 14.3 shows many probability scores less than **0.9**, and if we set the threshold for *don't know* at **0.9**, a *don't know* answer will be very frequent. As you will recall from *Figure 14.1*, this class was also much smaller than the other classes, which probably accounts for these low scores. Clearly, the DESC class will be problematic in deployment.

A confusion matrix, such as the one we reviewed in *Chapter 13*, can also help detect underperforming classes. We can generate a confusion matrix for the TREC data with the following code:

```
y_pred = classifier_model.predict(x_test)
y_pred = np.where(y_pred > .5, 1,0)
print(y_pred)
print(y_test)

predicted_classes = []
```

```
for i in range(len(y_pred)):
    max_index = np.argmax(y_pred[i])
    predicted_classes.append(max_index)
# View the results as a confusion matrix
from sklearn.metrics import confusion_
matrix,ConfusionMatrixDisplay,f1_score,classification_report
conf_matrix = confusion_matrix(y_test,predicted_classes,
    normalize=None)
```

This code generates the predicted classes from the test data (represented in the predicted_classes variable) and compares them to the true classes (represented in the y-test variable). We can use the scikit-learn confusion_matrix function to display the confusion matrix as follows:

```
# Displaying the confusion matrix
import matplotlib.pyplot as plt
plt.rcParams.update({'font.size': 12})

disp = ConfusionMatrixDisplay(confusion_matrix =
    conf_matrix, display_labels = class_names)
print(class_names)
disp.plot(xticks_rotation=75,cmap=plt.cm.Blues)
plt.show()
```

We can see the resulting confusion matrix in *Figure 14.4*. The confusion matrix tells us how often each class was predicted to be each other class, including itself:

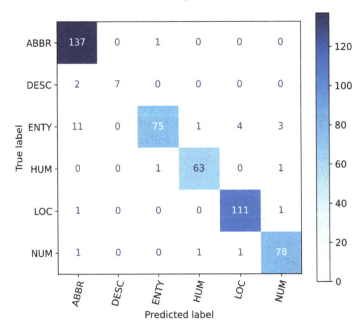

Figure 14.4 – The confusion matrix for the TREC test set

The correct predictions can be seen on the main diagonal. For example, ABBR was correctly predicted as ABBR *137* times. We can also see the prediction errors for each class. The most frequent error was incorrectly classifying ENTY as ABBR *11* times. In this particular example, we don't see a lot of evidence that specific classes get confused with each other, although there is a tendency for ENTY to be confused with ABBR.

Finally, we can look at the classification report to see the `precision`, `recall`, and `F1` scores for each class, as well as the overall averages for the entire test set. The recall scores in the classification report for DESC and ENTY are somewhat lower than the other recall scores, which reflects the fact that some of the items in those classes are incorrectly recognized as ABBR:

```
print(classification_report(y_test, predicted_classes, target_names =
class_names))
['ABBR', 'DESC', 'ENTY', 'HUM', 'LOC', 'NUM']
              precision    recall  f1-score   support

        ABBR       0.90      0.99      0.94       138
        DESC       1.00      0.78      0.88         9
        ENTY       0.97      0.80      0.88        94
         HUM       0.97      0.97      0.97        65
         LOC       0.96      0.98      0.97       113
         NUM       0.94      0.96      0.95        81

    accuracy                           0.94       500
   macro avg       0.96      0.91      0.93       500
weighted avg       0.94      0.94      0.94       500
```

It's worth pointing out at this point that the decision of whether the system is *good enough* really depends on the application and the developer's decision. In some applications, it's better to give the user some result, even if it might be wrong, while in other applications, it's important for every result to be correct, even if the system has to say *I don't know* almost all the time. Going back to the ideas of precision and recall that we covered in *Chapter 13*, another way of putting this is to say that in some applications, recall is more important, and in other cases, precision is more important.

If we want to improve the performance of the TREC application, the next step is to decide how to address our performance concerns and improve overall accuracy.

Fixing accuracy problems

In this section, we will look at fixing performance problems through two strategies. The first one involves issues that can be addressed by changing data, and the second strategy involves issues that require restructuring the application. Generally, changing the data is easier, and it is a better strategy if it is important to keep the structure of the application the same – that is, we don't want to remove classes or introduce new classes. We'll start by discussing changing the data and then discuss restructuring the application.

Changing data

Changing data can greatly improve the performance of your system; however, you won't always have this option. For example, you might not have control over the dataset if you work with a standard dataset that you intend to compare to other researchers' work. You can't change the data if you are in that situation because if you do, your system's performance won't be comparable to that of other researchers. If your system's performance isn't satisfactory but you can't change the data, the only options are to improve the algorithms by using a different model or adjusting the hyperparameters.

On the other hand, if you work on an application where you do have control over a dataset, changing data can be a very effective way to improve your system.

Many performance issues are the result of not having enough data, either overall, or in specific classes. Other performance issues can be due to annotation errors . We'll start with a brief discussion of annotation errors.

Annotation errors

It is possible that the poor performance of systems in supervised learning applications is due to annotation errors. Another way of putting this is to say that the supervision of data was wrong, and the system was trained to do the wrong thing. Perhaps an annotator accidentally assigned some data to the wrong class. If the data is training data, data in the wrong class will make the model less accurate, or if the data is test data, the item would be scored incorrectly because the model was wrong.

Checking for occasional annotation errors by reviewing the annotation of every item in the dataset can be very time-consuming, and it is not likely to improve the system much. This is because if the dataset is large enough, this kind of sporadic error is unlikely to have much of an impact on the quality of the overall system. However, if you suspect that annotation errors are causing problems, a simple check for low-confidence items can be helpful without requiring every annotation to be checked. This can be done by using a variation of the code we used in the *Checking for weak classes* section to check for weak classes. In that code, we predicted the class of each item in the dataset, kept track of its probabilities (scores), and averaged the probabilities of all the items in the class. To modify the code to look instead for individual items with low probabilities, you could record each item and its probability individually, and then look for low-probability items in the final list. You are encouraged to try this exercise for yourself.

On the other hand, it is also possible that data contains not only occasional mistakes but also systematic annotation errors. Systematic errors might be due to differences in the annotators' understanding of the meanings of the classes, leading to the similar items being assigned to different classes by different annotators. Ideally, these kinds of errors can be avoided, or at least reduced, by preparing clear annotation guidelines for annotators before the annotation process begins, or even by giving them training classes.

Tools such as the *kappa* statistic, which was mentioned in *Chapter 5*, can measure divergent annotations among annotators. If the kappa statistic shows that there is a lot of divergence across annotators, some of the data might need to be re-annotated using clarified guidelines. It can also happen that it is impossible to get annotators to agree because the decisions that the annotators have to make are inherently too subjective for people to agree on, no matter how much guidance they are given. This is a sign that the problem is not really suitable for NLU in the first place because there might not be a real correct classification for this data.

However, assuming that we do have a problem with objective classifications, in addition to addressing annotation errors, we can also improve system performance by creating a more balanced dataset. To do this, we will first look at adding and removing existing data from classes.

Adding and removing existing data from classes

Unbalanced amounts of data in different classes are a common situation that can lead to poor model performance. The main reason that a dataset can be unbalanced is that this imbalance represents the actual situation in the application domain. For example, an application that is supposed to detect online hate speech will most likely encounter many more examples of non-hate speech than actual hate speech, but it is nevertheless important to find instances of hate speech, even if they are rare. Another example of a naturally unbalanced dataset would be a banking application where we find many more utterances about checking account balances than utterances about changing account addresses. Changing the address on an account just doesn't happen very often compared to checking balances.

There are several ways to make the sizes of the classes more even.

Two common approaches are to duplicate data in the smaller classes or remove data from the larger classes. Adding data is called **oversampling** and removing data is called **undersampling**. The obvious approach to oversampling is to randomly copy some of the data instances and add them to the training data. Similarly, you can undersample by randomly removing instances from the classes that are too large. There are also other more sophisticated approaches to undersampling and oversampling, and you can find many online discussions about these topics – here, for example: `https://www.kaggle.com/code/residentmario/undersampling-and-oversampling-imbalanced-data`. However, we will not review these here because they can become quite complex.

Undersampling and oversampling can be helpful, but you should understand that they have to be used thoughtfully. For example, in the TREC dataset, trying to undersample the five frequent classes so that they have no more instances than the `DESC` class would require throwing out hundreds of instances from the larger classes, along with the information that they contain. Similarly, oversampling a small class such as `DESC` so that it contains the same number of instances as the larger classes means that there will be many duplicate instances of the `DESC` texts. This could result in overfitting the examples in `DESC` and consequently make it hard for the model to generalize to new test data.

It is easy to see that while undersampling and oversampling can potentially be useful, they are not automatic solutions. They are probably most helpful when classes are not extremely different in size and where there are plenty of examples, even in the smallest classes. You should also keep in mind that the classes don't have to be exactly balanced for a system to perform well.

Another approach to adding data is to create new data, which we will discuss in the next section.

Generating new data

If your dataset has underrepresented classes, or is too small overall, you can also add generated data to the entire dataset or just to the smaller classes. We will look at the following three ways to do this:

- Generating new data from rules
- Generating new data from LLMs
- Using crowdworkers to get new data

Generating new data from rules

One way to create new data is to write rules to generate new examples of data, based on the data that you already have. The **Natural Language Toolkit** (**NLTK**), which we used in *Chapter 8*, can be useful for this. As an example, let's suppose you're working on a local business search chatbot, and find that you need more data in the restaurant search class. You could write a **context-free grammar** (**CFG**), which we covered in *Chapter 8*, to generate more data with the following code that uses the NLTK parse library:

```
from nltk.parse.generate import generate
from nltk import CFG
grammar = CFG.fromstring("""
S -> SNP VP
SNP -> Pro
VP -> V NP PP
Pro -> 'I'
NP -> Det Adj N
Det -> 'a'
N -> 'restaurant' | 'place'
V -> 'am looking for' | 'would like to find'
PP -> P Adv
P -> 'near'| 'around'
Adv -> 'here'
Adj -> 'Japanese' | 'Chinese' | 'Middle Eastern' | 'Mexican'
for sentence in generate(grammar,n = 10):
    print(" ".join(sentence))
```

Note that the rules in an NLTK CFG can be any context-free rules; they don't have to correspond to actual linguistic categories. For example, we could have called the last Adj_Cuisine instead. We might want to do this if we want to be able to generate sentences with other adjectives, such as good or low-priced. The rule names and the rules themselves don't matter to the NLTK CFG package; the only thing that matters is that the CFG is written in the syntax that the NLTK CFG package expects. The names and the rules can be any rules that you find convenient to generate new examples.

The last two lines in the preceding code will generate 10 examples of sentences from this grammar, with the following result:

```
I am looking for a Japanese restaurant near here
I am looking for a Japanese restaurant around here
I am looking for a Japanese place near here
I am looking for a Japanese place around here
I am looking for a Chinese restaurant near here
I am looking for a Chinese restaurant around here
I am looking for a Chinese place near here
I am looking for a Chinese place around here
I am looking for a Middle Eastern restaurant near here
I am looking for a Middle Eastern restaurant around here
```

If you want to generate all of the possible sentences from these rules, you will leave out the parameter, n=10.

This is a fast way to generate a lot of sentences, but as you can see, the sentences are quite repetitious. This is because the NLTK `generate` method will produce every possible sentence that the grammar covers. Adding a lot of repetitious sentences to your training set could skew the model to these kinds of sentences, which in turn might make it harder for the model to recognize more varied restaurant search sentences. One approach to getting a wider variety of sentences from an NLTK CFG would be to write a broader grammar, generate all the sentences it covers, and then randomly select a subset of the generated sentences to add to the training set.

Using **large language models** (**LLMs**) to generate new examples is another useful and easy option, which we will discuss in the following section.

Generating new data from LLMs

Online LLMs such as ChatGPT are another very good way to get more training data because you can simply ask them to generate the appropriate training data. For example, let's say ChatGPT was given the following prompt:

generate 20 requests to find local restaurants of different cuisines and price ranges

ChatGPT (`chat.openai.com/chat`) would produce the following answer:

generate 20 requests to find local restaurants of different cuisines and price ranges

1. Find local Italian restaurants with a moderate price range.
2. Search for nearby sushi restaurants with a high-end price range.
3. Discover Mexican restaurants in the area with an affordable price range.
4. Locate Indian restaurants nearby with a mid-range price.
5. Find Chinese restaurants in the vicinity with a budget-friendly price range.
6. Search for Thai restaurants in the area with a high-end price range.

Figure 14.5 – The ChatGPT-generated restaurant query data

(For brevity, not all the results are shown.)

You can see that these sentences are much less repetitious than the ones from NLTK's `generate` method. In the initial ChatGPT prompt, you can also restrict the question style – for example, you could ask for generated questions in an informal style. This results in informal sentences such as the following:

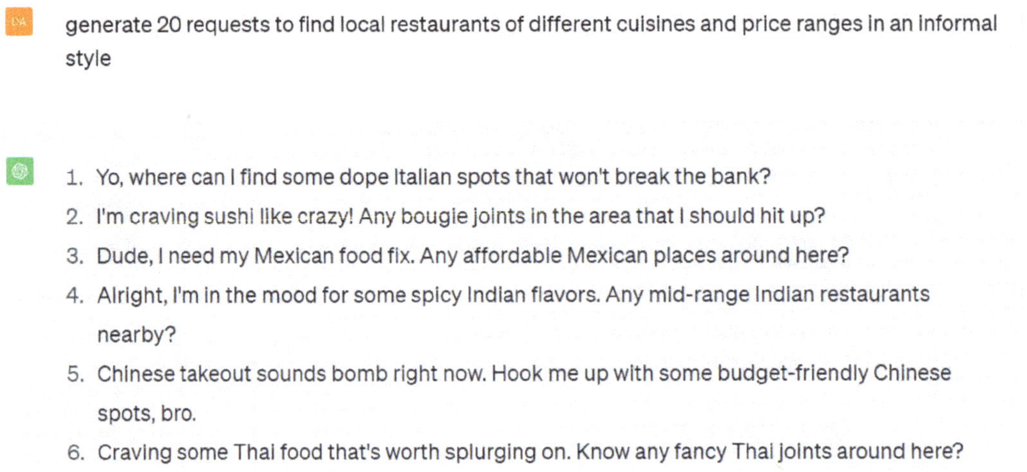

generate 20 requests to find local restaurants of different cuisines and price ranges in an informal style

1. Yo, where can I find some dope Italian spots that won't break the bank?
2. I'm craving sushi like crazy! Any bougie joints in the area that I should hit up?
3. Dude, I need my Mexican food fix. Any affordable Mexican places around here?
4. Alright, I'm in the mood for some spicy Indian flavors. Any mid-range Indian restaurants nearby?
5. Chinese takeout sounds bomb right now. Hook me up with some budget-friendly Chinese spots, bro.
6. Craving some Thai food that's worth splurging on. Know any fancy Thai joints around here?

Figure 14.6 – ChatGPT-generated restaurant query data in an informal style

You can also control the variation among the responses by changing the *temperature* parameter, which is available in the ChatGPT API. Temperature settings vary between zero and two. A temperature setting of zero means that the responses will be less varied, and a higher setting means that they will be more varied.

Within ChatGPT, a low temperature means that the generation model chooses the next word for a response among the higher probability words, and a high temperature means that the model will select the next word from among the words with lower probabilities. The result with a higher temperature setting will include more varied responses, but some of them might not make sense. *Figure 14.7* shows how to set the temperature in code directly from the API.

```
 1  import os
 2  import openai
 3  openai.api_key = OPENAIKEY
 4
 5  completion = openai.ChatCompletion.create(
 6      model="gpt-3.5-turbo",
 7      temperature=1.5,
 8      messages=[
 9          {"role": "user", "content": "generate 6 ways of asking for local restaurants of different cusines and price ranges"}
10      ]
11  )
12
13  result = completion.choices[0].message
14  result_content = result.get("content")
15  print(result_content)
16

1. "Can you recommend a good Italian restaurant that's not too expensive around here?"
2. "Are there any Thai restaurants that you would suggest trying in this area?"
3. "Could you point us in the direction of a steakhouse in town that won't break the bank?"
4. "We're looking for a Mexican restaurant with authentic cuisine. Any suggestions?"
5. "Do you know of any seafood restaurants that are a reasonable price nearby?"
6. "Where would you recommend going for some upscale dining? Possibly French cuisine."
```

Figure 14.7 – Setting the GPT temperature using the OpenAI API

The code in *Figure 14.7* sets the value of the `temperature` parameter to `1.5`, which results in a fairly diverse set of responses. You can see these at the bottom of *Figure 14.7*. The code also sets the `model` parameter to use the `gpt-3.5-turbo` model and sets the message to send in the `messages` parameter. If you are interested in experimenting with other GPT API calls, you can find other API parameters in the OpenAI API documentation at `https://platform.openai.com/docs/api-reference`.

Note that you will need to set the `openai.api_key` variable at line 3 to your own OpenAI user key to run this code, since the OpenAI API is a paid service.

If you use an LLM to generate data, be sure to check the results and decide whether the responses represent the kind of examples that your users would really say and, hence, should be included in your training data. For example, some of the informal requests in *Figure 14.6* might be more informal than many users would say to a chatbot.

A final way to add new data to underrepresented classes is to hire crowdworkers to create more data.

Using crowdworkers to get new data

Getting more data from crowdworkers is time-consuming and possibly expensive, depending on how much data you need and how complicated it is. Nevertheless, it would be an option if you didn't get enough data using other methods.

Of course, all of these methods we have outlined in this section (using rules, using LLMs, and using crowdworkers) can be combined – not all of the new training data has to come from the same place.

Another approach similar to changing the data is to change the application itself, which we will discuss in the next section.

Restructuring an application

In some cases, the best solution to classes that are not being predicted well is to restructure an application. As in the case of changing the data, you won't always have the option to do this if this is a standard dataset that the research community uses to compare work among different labs, as the application has to have the same structure as that used by other researchers for the results to be comparable.

If you do have control over the application structure, you can add, remove, or combine classes that don't perform well. This can greatly improve the overall application performance. Let's start by looking at an artificial example of an application that needs restructuring, and the different ways that this restructuring might be done.

Visualizing the need for class restructuring

Visualizing datasets can often provide immediate insight into potential performance problems. We can visualize how similar classes in a dataset are to each other in a couple of ways.

First, confusion matrices such as the one in *Figure 14.4* are a good source of information about which classes are similar to each other and consequently get confused for each other. We saw immediately from *Figure 14.4* that ENTY and DESC were quite often confused with ABBR. We might want to add data to those classes, as discussed in the previous section, or we could also consider restructuring the application, which we will discuss next.

A second visualization technique is to use the topic modeling techniques that we saw in *Chapter 12*, to see problems with the application structure.

Figure 14.8 shows how an artificially constructed dataset of four classes might look if we clustered them based on the *Chapter 12* tools, **Sentence Bert** and **BERTopic**. We can immediately see that there are some problems with the classes in this dataset.

Figure 14.8 – Unsupervised clustering of four classes with artificially generated data

First of all, the instances of **Class 1**, represented by circles, seem to cluster into two different classes, one centered around the point (*0.5, 0.5*) and the other centered around the point (*0.5, 1.75*). It seems unlikely that these clusters should both be grouped into the same class if they are actually that different. **Class 1** should probably be split, and the instances currently assigned to Class 1 should be assigned to at least two, and possibly three, new classes.

Class 2, represented by squares, and **Class 3**, represented by triangles, seem problematic. They are not completely mixed together, but they are not completely separate either. Some of the instances of both classes are likely to be misclassified because of their similarity to the other class. If you see classes such as **Class 3** and **Class 4**, with this kind of overlap, consider merging the classes if they appear to be similar in meaning (if they aren't similar in meaning, consider adding more data to one or both classes). Finally, **Class 4**, represented by stars, is compact and doesn't overlap with any other classes. It shouldn't require any adjustments.

Let's now take a look at three restructuring options – merging classes, dividing classes, and introducing new classes.

Merging classes

Classes that are much smaller than the rest of the classes can be merged with other semantically similar classes, especially if they are frequently confused with those classes. This can be a good strategy because, in many real applications, there are classes that simply don't occur very often, but unlike the hate speech example mentioned earlier, it isn't always critical to be able to tell the difference between the original classes. Of course, this is only possible with a multi-class problem – that is, a problem with more than two classes – since merging the classes in a binary (two-class) problem will put everything in one class and leave us with nothing to classify.

In some cases, merging classes can be accomplished by adding all the data from one class to the data of the other class, which is the simplest restructuring that we can do. A slightly more complex merger of classes can be done if the new structure is more complicated – for example, if it involves adding slots. For example, it is possible that classes such as **Class 2** and **Class 3** in *Figure 14.8* are actually not different enough to be worth trying to separate.

As an example, suppose we work on a generic personal assistant application, with classes such as `play music`, `find a restaurant`, `get weather forecast`, `find a bookstore`, and `find a bank`. It might turn out that `find a restaurant` has much more data than `find a bookstore`, and as a result, `find a bookstore` is often confused with `find a restaurant`. In that case, it would be worth considering whether all the `find a` classes should be merged into one larger class. This class could be called `local business search`, with `bookstore`, `restaurant`, and `bank` being treated as slots, as discussed in *Chapter 9*.

Another strategy is to separate classes such as **Class 1** in *Figure 14.5* into two different classes. The next section discusses dividing classes.

Dividing classes

Sometimes it makes sense to divide a large class into separate smaller classes if there appear to be systematic differences between two or more groups within the class. Tools such as BERTopic can help suggest names for new classes if the new name isn't obvious from looking at the instances in each group. Unfortunately, dividing a class into new classes isn't as easy as merging classes because the examples in the new classes will need to be annotated with their new names. Although re-annotation is more work, dividing and re-annotating classes is necessary if you have to divide a large class into more meaningful new classes.

Introducing an "other" class

Introducing an `other` class is a variant of the strategy of merging classes. If there are several small classes that don't really have enough training data to be reliably classified, it can sometimes be useful to group them together in an `other` class – that is, a class that contains items that don't fit into any of the other classes. One type of application that this approach can be useful for is call routing in a telephone self-service application.

In these applications, there can sometimes be hundreds of destinations where a call can be routed. In nearly every application of this kind, there are some infrequent classes for which there is much less data than other classes. Sometimes, it is best to not try to identify these classes because trying to do so accurately will be difficult with the small amounts of data available. A better strategy would be to group them together into an `other` class. It still might be hard to identify items in the `other` category because the items it contains will not be very similar, but it will keep them from interfering with the overall application accuracy. How items in the `other` class are handled depends on the specific application's goals, but options include handling them manually (for example, with a human call center agent) or simply telling users that the system can't handle their question.

After the accuracy issues that were identified during initial development have been identified and addressed, it is time to deploy the system.

Moving on to deployment

If we've fixed the performance issues we've discussed so far, we will have trained a model that meets our performance expectations, and we can move on to deployment, when the system is installed and does the task that it was designed for. Like any software, a deployed NLU model can have problems with system and hardware issues, such as network issues, scalability, and general software problems. We won't discuss these kinds of problems because they aren't specific to NLU.

The next section will cover considerations to address NLU performance problems that occur after deployment.

Problems after deployment

After an NLU system is developed and put into place in an application, it still requires monitoring. Once the system has reached an acceptable level of performance and has been deployed, it can be tempting to leave it alone and assume that it doesn't need any more attention, but this is not the case. At the very least, the deployed system will receive a continuous stream of new data that can be challenging to the existing system if it is different from the training data in some way. On the other hand, if it is not different, it can be used as new training data. Clearly, it is better to detect performance problems from internal testing than to learn about them from negative customer feedback.

At a high level, we can think of new performance problems as either being due to a change in the system itself, or due to a change in the deployment context.

Changes in system performance due to system changes should be detected by testing before the new system is deployed. This kind of testing is very similar to the kind of testing that has to be done for any software deployment, so we won't cover it in any detail. Degradation in performance can be detected by versioning the system and running an evaluation with a fixed set of data and metrics after every change. This is useful to both detect decreases in performance but also to document improvements in performance.

As with any machine-learning-based system, new data can cause problems with an NLU system because it is different in some significant way from the training data. These kinds of differences are frequently due to changes in the deployment context.

What do we mean by changes in the **deployment context**? The deployment context refers to everything about the application, except for the NLU system itself. Specifically, it can include the users, their demographics, their geographical locations, the backend information that's being provided, and even events in the world such as weather. Any of these can change the characteristics of texts that the application processes. These changes alter the correspondence between the training data and the new data being processed, which will lead to a decrease in performance.

Some changes in the deployment context can be predicted. For example, if a company introduces a new product, this will introduce new vocabulary that a customer support chatbot, voice assistant, or email router needs to recognize, since, after all, we expect customers to be talking about it. It is a best practice to perform an evaluation on new data after changes like the introduction of a new product occurs, and decide whether the system should be retrained with additional data.

On the other hand, some changes can't be predicted – for example, the COVID-19 pandemic introduced a lot of new vocabulary and concepts that medical or public health NLU applications needed to be trained on. Because some deployment context changes can't be predicted, it is a good idea to periodically perform an evaluation using new data coming in from the deployment.

Summary

In this chapter, you have learned about a number of important strategies to improve the performance of NLU applications. You first learned how to do an initial survey of the data and identify possible problems with the training data. Then, you learned how to find and diagnose problems with accuracy. We then described different strategies to improve performance – specifically, adding data and restructuring the application. The final topic we covered was a review of problems that can occur in deployed applications and how they can be addressed.

In the final chapter, we will provide an overview of the book and a look to the future. We will discuss where there is potential for improvement in the state of the art of NLU performance, as well as faster training, more challenging applications, and what we can expect from NLU technology as the new LLMs become more widely used.

15

Summary and Looking to the Future

In this chapter, we will get an overview of the book and a look into the future. We will discuss where there is potential for improvement in performance as well as faster training, more challenging applications, and future directions for practical systems and research.

We will cover the following topics in this chapter:

- Overview of the book
- Potential for better accuracy and faster training
- Other areas for improvement
- Applications that are beyond the current state of the art
- Future directions

The first section of this chapter is an overall summary of the topics covered in this book.

Overview of the book

This book has covered the basics of **natural language understanding** (NLU), the technology that enables computers to process natural language and apply the results to a wide variety of practical applications.

The goal of this book has been to provide a solid grounding in NLU using the Python programming language. This grounding will enable you not only to select the right tools and software libraries for developing your own applications but will also provide you with the background you need to independently make use of the many resources available on the internet. You can use these resources to expand your knowledge and skills as you take on more advanced projects and to keep up with the many new tools that are becoming available as this rapidly advancing technology continues to improve.

In this book, we've discussed three major topics:

- In *Part 1*, we covered background information and how to get started

- In *Part 2*, we went over Python tools and techniques for accomplishing NLU tasks

- In *Part 3*, we discussed some practical considerations having to do with managing deployed applications

Throughout this book, we have taken a step-by-step approach through the typical stages of an NLU project, starting from initial ideas through development, testing, and finally, to fine-tuning a deployed application. We can see these steps graphically in *Figure 15.1*:

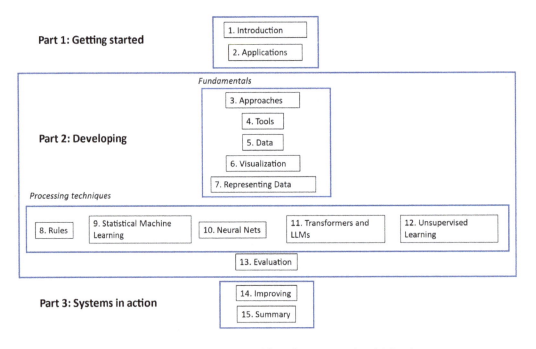

Figure 15.1 – The NLU project life cycle as covered in this book

In *Part 1*, you were introduced to the general topic of NLU and the kinds of tasks to which it can be applied.

In *Part 2*, we started by covering foundational topics that support the most successful NLU applications, such as software development tools, data, visualization, and approaches to representing NLU data. The second general topic that we covered in *Part 2* was a set of five different approaches to processing language, including rules, traditional machine learning techniques, neural networks, transformers, and unsupervised learning. These topics are covered in the five chapters from *Chapter 8* through *Chapter 12*, which covered the basics of NLU algorithms. Mastering this material will give you the background that you need to continue exploring NLU algorithms beyond the topics covered in this book. The final topic in *Part 2* was the very important subject of evaluation. Evaluation is critically important both for practical NLU deployments and for successful academic research. Our review of evaluation in *Chapter 13*, covers a variety of the most important NLU evaluation tools. With this background, you should now be prepared to evaluate your own NLU projects.

Finally, in *Part 3*, we turned to the topic of systems in action and focused (particularly in *Chapter 14* on approaches to improving the performance of systems both before and after deployment.

If you continue to work in the field of NLU, you will find that there are still many challenges, despite the fact that recent advances in **large language models** (**LLMs**) have dramatically improved the performance of NLU systems on many tasks.

We will look at some of these challenges in the next two sections, starting with the most important areas of improvement – better accuracy and faster training – followed by a section on other areas of improvement.

Potential for improvement – better accuracy and faster training

At the beginning of *Chapter 13*, we listed several criteria that can be used to evaluate NLU systems. The one that we usually think of first is accuracy – that is, given a specific input, did the system provide the right answer? Although in a particular application, we eventually may decide to give another criterion priority over accuracy, accuracy is essential.

Better accuracy

As we saw in *Chapter 13*, even our best-performing system, the large **Bidirectional Encoder Representations from Transformers** (**BERT**) model, only achieved an F_1 score of *0.85* on the movie review dataset, meaning that 15% of its classifications were incorrect. State-of-the-art LLM-based research systems currently report an accuracy of *0.93* on this dataset, which still means that the system makes many errors (SiYu Ding, Junyuan Shang, Shuohuan Wang, Yu Sun, Hao Tian, Hua Wu, and Haifeng Wang. 2021. *ERNIE-Doc: A Retrospective Long-Document Modeling Transformer*), so we can see that there is still much room for accuracy improvements.

LLMs represent the state of the art in NLU. However, there have been few formal studies of the accuracy of the most recent LLMs, so it is difficult to quantify how good they are. One study of a fairly challenging medical information task where physicians evaluated the accuracy of ChatGPT's answers found that ChatGPT answers were considered to be largely accurate overall, receiving a mean score of *4.6 out of 6* (Johnson, D., et al. (2023). *Assessing the Accuracy and Reliability of AI-Generated Medical Responses: An Evaluation of the Chat-GPT Model*. Research Square, rs.3.rs-2566942. `https://doi.org/10.21203/rs.3.rs-2566942/v1`). However, the system still made many errors, and the authors cautioned that it was important for physicians to review medical advice supplied by ChatGPT or, in general, any LLMs at the current state of the art.

Better accuracy will always be a goal in NLU. Achieving better accuracy in future systems will include developing larger pretrained models as well as developing more effective fine-tuning techniques. In addition, there is a significant amount of work to be done in extending the current high performance of LLMs for the most widely studied languages to less well-studied languages.

Faster training

By working through some of the exercises in earlier chapters, you have found that training NLU models from scratch or fine-tuning an LLM did not take more than a few hours of computer time. For the purposes of this book, we intentionally selected smaller datasets so that you could get feedback from the training process more quickly. However, even larger problems that you may want to address in a practical setting should not take more than a few days of training time. On the other hand, training pretrained LLMs can take a very long time. One estimate for the training time for GPT-3 was that it took 355 GPU years to train on the 300 billion token training dataset. In practice, the calendar time required was reduced by running the training on multiple GPUs in parallel (`https://lambdalabs.com/blog/demystifying-gpt-3`). Still, this training does involve tremendous amounts of computing power, along with the associated costs.

Since most pretrained models are trained by large organizations with extensive computing resources rather than by smaller organizations or researchers, long training times for large models don't directly affect most of us because we will be using the pretrained models developed by large organizations. However, these long training times do affect us indirectly because long training times mean that it will take longer for new and improved models to be released for general use.

In addition to better accuracy and faster training times, there are other areas where NLU technology can be improved. We will review some of these in the next section.

Other areas for improvement

The areas for improvement that we'll review in this section are primarily related to making NLU technology more practical in various ways, such as speeding up development and decreasing the number of computer resources needed during development and at runtime. These topics include smaller models, more explainability, and smaller amounts of fine-tuning data.

Smaller models

The BERT models we looked at in *Chapter 11* and *Chapter 13*, were relatively small. The reason for choosing these models was so that they could be downloaded and fine-tuned in a relatively short amount of time. However, as a rule of thumb, large models will be more accurate than smaller models. But we can't always take advantage of large models because some models are too large to be fine-tuned on a single GPU, as pointed out on the TensorFlow site (`https://colab.research.google.com/github/tensorflow/text/blob/master/docs/tutorials/classify_text_with_bert.ipynb#scrollTo=dX8Ft1pGJRE6`). Because the larger models have better accuracy, it would be very helpful if high-accuracy performance could be obtained with smaller models. In addition, there are many situations where large models will not fit on resource-constrained devices such as mobile phones or even smartwatches. For those reasons, decreasing the size of models is an important goal in NLU research.

Less data required for fine-tuning

For most NLU applications that use pretrained models, the pretrained model needs to be fine-tuned with application-specific data. We covered this process in *Chapter 11* and *Chapter 13*. Clearly, reducing the amount of data required to fine-tune the system results in a reduction in the development time for the fine-tuned system. For example, in its discussion of the process of fine-tuning GPT-3, OpenAI states, *"The more training examples you have, the better. We recommend having at least a couple hundred examples. In general, we've found that each doubling of the dataset size leads to a linear increase in model quality"* (https://platform.openai.com/docs/guides/fine-tuning). As we learned in *Chapter 5*, the process of finding and annotating data can be time-consuming and expensive, and it is clearly desirable to minimize this effort.

Explainability

For the most part, a result from an NLU system based on machine learning will simply be a number, such as a probability that a text falls into one of the categories on which the model was trained. We don't have an easy way to understand how the system came up with that answer, whether the answer is correct or incorrect. If the answer is incorrect, we can try to improve the model by adding more data, adjusting the hyperparameters, or using some of the other techniques that we reviewed in *Chapter 14*, but it's hard to understand exactly why the system came up with the wrong answer.

In contrast, if a rule-based system such as those we went over in *Chapter 8* makes an error, it can normally be traced back to an incorrect rule, which can be fixed. However, since nearly all current systems are based on machine learning approaches rather than rules, it is very difficult to understand how they arrive at the answers that they do. Nevertheless, it is often important for users to understand how a system came up with a result. If the users don't understand how the system came up with an answer, they might not trust the system. If a system gives a wrong answer or even a correct answer that the user doesn't understand, it can undermine user confidence in the system. For that reason, explainability in NLU and in AI, in general, is an important research topic. You can read more about this topic at https://en.wikipedia.org/wiki/Explainable_artificial_intelligence.

Timeliness of information

In *Chapter 14*, we discussed how changes in the deployment context can result in system errors. The introduction of new product names, new movies, or even significant news events can lead to the system not knowing the answer to a user's question or even giving the wrong answer. Because LLMs take so long to train, they are especially vulnerable to making errors due to a change in the deployment context.

For example, the ChatGPT system has a knowledge cutoff date of September 2021, which means that it doesn't have any knowledge of events that occurred after that. Because of this, it can make mistakes like the one shown in *Figure 15.2*, which states that the current monarch of the United Kingdom is Elizabeth II. This was true in September 2021, but it is no longer true.

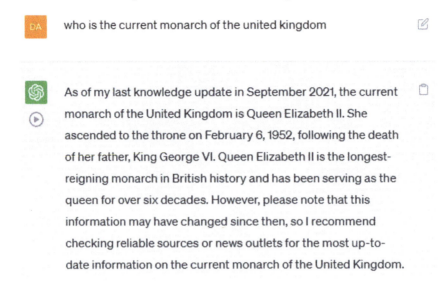

Figure 15.2 – ChatGPT answer to "who is the current monarch of the united kingdom"

Although the ChatGPT system acknowledges that its information may be out of date, this lack of timeliness can lead to errors if something changes in the deployment context. If you are developing your own application and something changes in the deployment context, you can either retrain the system from scratch with the new data or add new data to your existing model. However, if you are using a cloud-based LLM, you should be aware that the information it provides can be out of date. Note that this cutoff period can vary between different LLMs. For example, the Google Bard system was able to correctly answer the question in *Figure 15.2*.

If your application uses an LLM and requires access to accurate time-dependent information, you should verify that the system you're using is being kept up to date by its developers.

The next section talks about a few blue sky applications that may be possible in the future.

Applications that are beyond the current state of the art

This section talks about several applications that are not yet possible, but that are theoretically feasible. In some cases, they could probably be achieved if the right training data and computing resources were available. In other cases, they might require some new algorithmic insights. In all of these examples, it is very interesting to think about how these and other futuristic applications might be accomplished.

Processing very long documents

Current LLMs have relatively small limits on the length of documents (or prompts) they can process. For example, GPT-4 can only handle texts of up to 8,192 tokens (`https://platform.openai.com/docs/models/gpt-4`), which is around 16 single-spaced pages. Clearly, this means that many existing documents can't be fully analyzed with these cloud systems. If you are doing a typical classification task, you can train your own model, for example, with a **Term frequency-inverse document frequency** (TF-IDF) representation, but this is not possible with a pretrained model.

In that case, the documents can be as long as you like, but you will lose the advantages of LLMs. Research systems such as Longformer have been able to process much longer documents through more efficient use of computational resources. If you have a use case for processing long documents, it would be worth looking into some of these research systems.

Understanding and creating videos

To understand videos, a system would need to be able to interpret both the video and audio streams and relate them to each other. If the system learns someone's name in the early part of the video and that character appears in a later part of the video, it should be able to name the person based on recognizing the image of the character. It could then do tasks such as transcribing the script of a movie by simply watching it, complete with notations like "Character X smiles". This is not a very difficult task for humans, who are quite good at recognizing a person that they've seen before, but it would be very difficult for automated systems. While they are quite good at identifying people in images, they are less capable of identifying people in videos. In contrast to understanding videos, generating videos seems to be an easier task. For example, there are currently systems available that generate videos from text, such as a system developed by Meta (`https://ai.facebook.com/blog/generative-ai-text-to-video/`), although the videos don't yet look very good.

Interpreting and generating sign languages

One application of understanding videos would be to understand sign languages such as American Sign Language and translate them into spoken languages. Combined with the reverse process of translating spoken language into sign language, this kind of technology could greatly simplify communication between signers and speakers of spoken languages. There have been some exploratory studies of interpreting and generating sign languages.

For example, the work on `https://abdulhaim.github.io/6.S198-Assignments/final_project.html` describes an approach to interpreting Argentinian Sign Language using **Convolutional Neural Networks** (CNNs). Although this is an interesting proof of concept, it only works with 64 signs from Argentinian Sign Language. In fact, there are thousands of signs used in actual sign languages, so handling 64 signs is only a small demonstration of the possibility of automatically interpreting sign languages.

In addition, this research only used hand positions to recognize signs, while, in fact, signs are also distinguished by other body positions. More work needs to be done to demonstrate practical automatic sign language interpretation. This would be greatly aided by the availability of more sign language datasets.

Writing compelling fiction

If you have experimented with ChatGPT or other LLMs, you may have noticed that the writing style is rather bland and boring. This is because it's based on text from the internet and other existing sources and there is no way for it to be creative beyond the data that it is trained on. On the other hand, compelling fiction is unique and often contains insights and verbal images that have never appeared in writing before.

As an example, let's look at an excerpt from one of the great poems of the English language, To a Skylark, by Percy Bysshe Shelley, which can be seen in *Figure 15.3*:

To a Skylark

By Percy Bysshe Shelley

Hail to thee, blithe Spirit!

Bird thou never wert,

That from Heaven, or near it,

Pourest thy full heart

In profuse strains of unpremeditated art.

Higher still and higher

From the earth thou springest

Like a cloud of fire;

The blue deep thou wingest,

And singing still dost soar, and soaring ever singest.

Figure 15.3 – An excerpt from "To a Skylark", a poem by Percy Bysshe Shelley (1820)

This poem includes novel figures of speech such as the simile that compares a bird to *a cloud of fire* and uses the metaphor *the blue deep* for the sky, which are probably unique in literature.

Compare this to the poem generated by ChatGPT in *Figure 15.4*. When prompted to write a poem about a skylark flying in the sky, the result seems flat and unoriginal compared to the Shelley poem and includes cliches such as *boundless sky* and *ascends on high*.

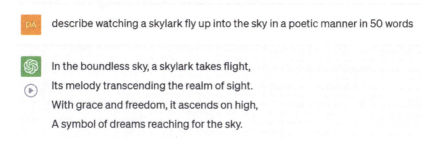

Figure 15.4 – ChatGPT poem about a skylark

Let's think about how we might train an LLM to learn how to generate good poetry or interesting fiction. If we follow the standard NLU development paradigm of learning a model from training data, for an NLU system to write compelling fiction, we would need a dataset consisting of text examples of compelling and engaging writing and other examples of writing that are not compelling or engaging. Alternatively, we might be able to identify other features of compelling writing (use verbs, avoid passive voice, etc.) that could be used to train systems to evaluate writing or to produce good writing. You can understand how far we are from this kind of application by thinking about what an NLU system would need to be able to do to write insightful book reviews. It would have to be familiar with the author's other books, other books in a similar genre, any relevant historical events mentioned in the book, and even the author's biography. Then it would have to be able to pull all this knowledge together into a concise analysis of the book. All of this seems quite difficult.

The next section will look at applications that are targets of current research and are much closer to realization than the ones we've just reviewed. We are likely to see advances in these kinds of applications in the next couple of years.

Future directions in NLU technology and research

While the recent improvements in NLU technology based on transformers and LLMs, which we reviewed in *Chapter 11*, have resulted in very impressive capabilities, it is important to point out that there are many topics in NLU that are far from solved. In this section, we will look at some of the most active research areas – extending NLU to new languages, speech-to-speech translation, multimodal interaction, and avoiding bias.

Quickly extending NLU technologies to new languages

A precise count of the number of currently spoken languages is difficult to obtain. However, according to *WorldData.info*, there are currently about 6,500 languages spoken throughout the world (`https://www.worlddata.info/languages/index.php#:~:text=There%20are%20currently%20around%206%2C500,of%20Asia%2C%20Australia%20and%20Oceania`). Some languages, such as Mandarin, English, Spanish, and Hindi, are spoken by many millions of people, while other languages are spoken by very few people and these languages are even in danger of dying out (for example, you can see a list of the endangered languages of North America alone at `https://en.wikipedia.org/wiki/List_of_endangered_languages_in_North_America`).

Languages with many millions of speakers tend to be more economically important than languages with few speakers, and as a consequence, NLU technology for those languages is generally much more advanced than that for languages with few speakers. If you recall from our discussion of LLMs in *Chapter 11*, training an LLM such as BERT or GPT-3 for one language is a very expensive and time-consuming process that requires enormous amounts of text data. It would be impractical for this training process to be carried out for thousands of languages. For that reason, researchers have looked into adapting LLMs used for widely spoken languages to less widely spoken languages.

This is a very active research area that presents many challenges to NLU technology. One challenge, for example, is how to keep the original language from being forgotten when its language model is adapted to a new language – a process called **catastrophic forgetting**.

Citation

This is an example of a recent research paper on this topic that you can read for more insight into the problem of adapting LLMs to new languages:

Cahyawijaya, S., Lovenia, H., Yu, T., Chung, W., & Fung, P. (2023). *Instruct-Align: Teaching Novel Languages with to LLMs through Alignment-based Cross-Lingual Instruction.* arXiv preprint arXiv:2305.13627. `https://arxiv.org/abs/2305.13627`.

Real-time speech-to-speech translation

Anyone who has traveled to a foreign country whose language they do not know, or whose language they do not know well, has probably found communication very frustrating. Looking up words or even phrases in handheld apps or paper dictionaries is slow and can be inaccurate. A much better solution would be speech-to-speech translation. Speech-to-speech translation technology listens to speech in one language, translates it to another language, and the system speaks the output in a second language, which would be much faster than typing words into a mobile app.

The base technologies underlying **speech-to-speech translation** are actually fairly advanced. For example, Microsoft Cognitive Services offers a speech-to-speech translation service (`https://azure.microsoft.com/en-us/products/cognitive-services/speech-translation/`) with support for over 30 languages. The number of available language pairs continues to increase – for example, Speechmatics offers a translation service for 69 language pairs (`https://www.speechmatics.com/product/translation`).

However, most of these services do their processing in the cloud. Because one of the most important use cases for speech-to-speech translation is for travel, users may not want to use a service that requires access to the cloud. They may not have a good internet connection, or they may not want to pay for data while traveling. It is much more difficult to translate speech offline, without sending it to the cloud, as mobile devices have far fewer computing resources than the cloud. The results are less accurate and not nearly as many languages are supported. For example, the Apple Translate app (`https://apps.apple.com/app/translate/id1514844618`) claims to support 30 languages but the reviews are very low, especially for offline use. There is significant room for improvement in the technology for offline speech-to-speech translation.

Multimodal interaction

Multimodal interaction is a type of user-system interaction where the user interacts with a computer system in multiple ways (*modalities*) in addition to language. For example, multimodal interaction could include camera input that allows the system to interpret facial expressions in addition to speech input. This would let the system read the user's body language to detect emotions such as happiness or confusion in addition to interpreting what the user says. As well as understanding multimodal user inputs, a multimodal system can also produce images, animations, videos, and graphics in addition to language in order to respond to users' questions.

> ### Citation
>
> Multimodal interaction has been used extensively in research projects such as the following:
>
> António Teixeira, Annika Hämäläinen, Jairo Avelar, Nuno Almeida, Géza Németh, Tibor Fegyó, Csaba Zainkó, Tamás Csapó, Bálint Tóth, André Oliveira, Miguel Sales Dias, *Speech-centric Multimodal Interaction for Easy-to-access Online Services – A Personal Life Assistant for the Elderly*, Procedia Computer Science, Volume 27, 2014, Pages 389-397, ISSN 1877-0509, `https://doi.org/10.1016/j.procs.2014.02.043`.

However, multimodal interaction is far from widespread in practical applications. This may be due in part to the relative scarcity of training data for multimodal systems since training multimodal systems requires data for all the modalities being used in the system, not just language data. For example, if we wanted to develop an application that used a combination of facial expression recognition and NLU to understand users' emotions, we would need a video dataset annotated with both facial expressions and NLU categories. There are a few existing datasets with this kind of information – for example, the datasets listed at `https://www.datatang.ai/news/60` – but they are not nearly as abundant as the text datasets that we've been working with throughout this book. Multimodal interaction is a very interesting topic, and the availability of additional data will certainly stimulate some future groundbreaking work.

Detecting and correcting bias

The training data for LLMs is based on existing text, primarily from the web. This text, in many cases, reflects cultural biases that we would not like to perpetuate in our NLU systems. It is easy to find this bias in current LLMs. For example, the article *ChatGPT insists that doctors are male and nurses female*, by Suzanne Wertheim, shows many examples of ChatGPT assuming that people in certain professions are male or female (`https://www.worthwhileconsulting.com/read-watch-listen/chatgpt-insists-that-doctors-are-male-and-nurses-female`). This problem has been the topic of considerable research and is far from solved.

> **Citation**
>
> You can find out more about how bias has been addressed in the following survey article:
>
> Alfonso, L. (2021). *A Survey on Bias in Deep NLP*. Applied Sciences, 11(7), 3184. `https://doi.org/10.3390/app11073184`.

Summary

In this chapter, we have summarized the previous chapters in the book, reviewed some areas where NLU technology still faces challenges, and talked about some directions where it could improve in the future. NLU is an extremely dynamic and fast-moving field, and it will clearly continue to develop in many exciting directions. With this book, you have received foundational information about NLU that will enable you to decide not only how to build NLU systems for your current applications but also to take advantage of technological advances as NLU continues to evolve. I hope you will be able to build on the information in this book to create innovative and useful applications that use NLU to solve future practical as well as scientific problems.

Further reading

SiYu Ding, Junyuan Shang, Shuohuan Wang, Yu Sun, Hao Tian, Hua Wu, and Haifeng Wang. 2021. *ERNIE-Doc: A Retrospective Long-Document Modeling Transformer*. In Proceedings of the 59th Annual Meeting of the Association for Computational Linguistics and the 11th International Joint Conference on Natural Language Processing (Volume 1: Long Papers), pages 2914–2927, Online. Association for Computational Linguistics

Beltagy, I., Peters, M.E., & Cohan, A. (2020). *Longformer: The Long-Document Transformer*. arXiv, abs/2004.05150

Index

A

ablation 232
accuracy 164, 236
accuracy problems
application, restructuring 270
changing data 264
fixing 263
activation function 181, 186
example 187
air travel information system (ATIS) 233
analytics 12
annotations 29, 30
APIsList 31
URL 31
Apple Translate app
reference link 285
application data 31
application programming interfaces (APIs) 20
application, restructuring 270
classes, dividing 272, 273
classes, merging 272
need, visualizing 270, 271
applications, without NLP 26
graphical interfaces, using 27, 28

inputs, recognizing from known list of words 27
sufficient data availability, ensuring 28, 29
text, analyzing with regular expressions 26
application-specific types, techniques 89
class labels, substituting for words and numbers 89, 90
data imbalance 91
domain-specific stopwords 90
HTML markup, removing 90
redaction 90
text preprocessing pipelines, using 91
area under the curve (AUC) 164, 237, 238
Artificial NNs (ANNs) 180
attention 197
applying, in transformers 198
automatic speech recognition (ASR) 6, 74

B

bag of words (BoW) 113-119, 131, 183, 217
batch 181
Bayesian classification 165
Beginning Inside Outside (BIO) 173
BERT model
compiling 205, 206
data, installing 201, 202

dataset, splitting into testing sets 202
dataset, splitting into training 202
dataset, splitting into validation 202
defining, for fine-tuning 203, 204
evaluating, on test data 209
loading 202, 203
loss function, defining 204
metrics, defining 204
number of epochs, defining 204, 205
optimizer, defining 204, 205
saving, for inference 209, 210
training 206
training process, plotting 207, 208
versus TF-IDF-Naïve Bayes
    classification 246
BERTopic 270
applying 219
document, finding 226
embeddings 219, 220
labeling 223
model, constructing 220-222
parameters 222
supervised classification application,
    developing 226
visualization 224, 225
**Bidirectional Encoder Representations
    from Transformers (BERT) 198, 277**
advantages 198
using 199, 200
variants 199
bigram 110
**bilingual evaluation understudy
    (BLEU) metric 232**
binary bag of words 131, 132
binary crossentropy 187
blue sky applications
compelling fiction, writing 282, 283
long documents, processing 281
sign languages, interpreting
    and generating 281
videos, creating 281

C

call centers
conversations 70
catastrophic forgetting 284
changing data 264
annotation errors 264
existing data, adding from classes 265
existing data, removing from classes 265
new data, generating 266
character encodings 5
ChatGPT
reference link 211
chat logs 71
classification 11, 44, 171
cloud-based LLMs 210
ChatGPT 211
GPT-3, applying 212-214
clustering 216, 217
Hierarchical Density-Based Spatial
    Clustering of Applications with
    Noise (HDBSCAN) 218
k-means algorithm 217
topic modeling, using 217
code-switching 26
color-hex
URL 159
computational efficiency, considerations
inference time 127, 128
training time 127
conditional random fields (CRF) 163
confusion matrix 238
constituency rules 152

context-free grammars (CFGs) 152-155

context-independent vectors

used, for representing words 137

conversational artificial intelligence 6

Convolutional Neural Networks (CNNs) 179, 192, 281

example 192

coreference 41

corpora

used, for generating data 74, 75

corpus 30

corpus properties 63-65

POS frequencies 65-67

count bag of words 132, 133

criteria, for evaluating NLU systems 277

better accuracy 277

explainability 279

faster training 278

less data, for fine-tuning 279

smaller models 278, 279

timeliness of information 279, 280

cross-entropy 204

cross-lingual optimized metric for evaluation of translation (COMET) 232

crowdworkers

treating, ethically 76

D

data

availability, ensuring 28, 29

crowdworkers, treating ethically 76

human subjects, treating ethically 76

non-text, removing 77, 78

obtaining, with pre-existing corpora 74, 75

preprocessing 77

sources, annotating 69, 70

sources, searching 69, 70

spelling correction 88

databases 71

data exploration 97

bag of words (BoW) 113,-119

documents similarities, measuring 113

frequency distributions 97

k-means clustering 113-119

data exploration, frequency distributions

data, visualizing with Matplotlib 103-105

data, visualizing with pandas 103-105

data, visualizing with Seaborn 103-105

frequencies, in corpus 110-113

positive movie reviews, versus negative movie reviews 106-109

word clouds 105, 106

word frequency distributions 98-103

data, for application

chat logs 71

conversations, in call centers 70

customer reviews 71

databases 71

message boards 71

searching 70

subsections 70

data, for research project

APIs 72

collecting 72

crowdsourcing 72

searching 71

Wizard of Oz (WoZ) methods 72

data imbalance 91

data, non-text

emojis, removing 78, 79

smart quotes, removing 79

data partitioning 233, 234

advantage 235

data representations
  working with 217
dataset 30
data, spelling correction
  contractions, expanding 89
data, text
  lemmatizing 84, 85
  lower casing 82, 83
  part of speech tagging 84, 85
  punctuation, removing 87
  regularizing 80
  stemming 83, 84
  stopword removal 86, 87
  tokenization 80-82
data visualization
  Sentence Polarity Dataset 96, 97
  significance 96
  text document dataset 96
  with Matplotlib 103-105
  with pandas 103-105
  with Seaborn 103-105
datetime package
  reference link 148
deep learning (DL) 57
  approaches 44
deep neural networks (DNN) 58
dependencies 152
dependency parsing 39
dependent variable 120
deployment context
  changes 274
development costs
  considering, for natural language
    applications 31
development test data 234
dialog management 6
dimensionality reduction 123

displaCy
  URL 59
document classification 11
document retrieval 12
documents
  classifying, with Naïve Bayes 165
  classifying, with SVMs 169-171
  representing, with TF-IDF 165
domain-specific stopwords 90

E

education 9
embeddings 219, 220
enterprise assistants 8, 9
entities 7, 158
epochs 180
error rate 164, 235
evaluation
  considerations 164
  overview 164
evaluation metrics
  accuracy 235
  AUC 237, 238
  confusion matrix 238
  error rate 235
  F1 236, 237
  precision 236
  recall 236
  ROC 237, 238
evaluation metrics selection, concepts
  reliability 235
  validity 235
evaluation paradigms
  ablation 232, 233
  language output, evaluating 232
  shared tasks 233

system results, comparing on
      standard metrics 232
explainability
  reference link 279
exploratory data analysis (EDA) 97
**Extensible Markup Language
      (XML) tags 173**

F

F1 236, 237
fake news detection 12
few-shot learning 210
fine-tuning 45
fixed expressions 145
flowchart
  for deciding, on NLU applications 33, 34
forward propagation 182
frequently asked questions (FAQs) 11
fully connected NN (FCNN) 180, 182

G

gated recurrent units (GRUs) 192, 197
**General Architecture for Text
      Engineering (GATE)**
  URL 73
generation 126
**Generative Pre-trained
      Transformers (GPTs) 45**
generic voice assistants 8
Gensim
  URL 58
geopolitical entity (GPE) 57
GitHub 53
  URL 53

gold standard 29
Google Translate
  reference link 13
GPT-3
  applying 212-214
GPT-3 Language Model
  reference link 278
GPT-4
  reference link 281
grammar 39
graphical user interface
  using 27, 28
  versus natural language
      understanding (NLU) 28
ground truth 29

H

**Hierarchical Density-Based Spatial
      Clustering of Applications with
      Noise (HDBSCAN) 218**
HTML markup
  removing 90
Hugging Face
  reference link 30
  URL 75
human subjects
  treating, ethically 76
hypernym 149
hyperparameters 127
  modifying, to improve
      performance 190, 191
hyponym 149

I

idioms 113
imbalanced dataset 126
independent variable 120
inflectional morphemes 83
inflectional morphology 83
information extraction 13
inputs
    recognizing, from known list of words 27
intent 7
intent recognition 7
interactive applications 7
    education 9, 10
    enterprise assistants 8, 9
    generic voice assistants 8
    translation 9
interactive development
        environments (IDEs) 51
inter-annotator agreement 74
Internet Movie Database (IMDB) 96
inverse document frequency (IDF) 133
isa relationships 149

J

JavaScript Object Notation
        (JSON) format 172
JupyterLab 51, 52
    setting up 60-62
    using 60

K

Kaggle
    URL 75
kappa statistic 74

Keras
    using 57, 58
keyword spotting 27
k-fold cross-validation 234
k-means algorithm 217
k-means clustering 113-119

L

label derivation
    topic modeling, using 217
labeling 223
language
    identification 26
    representing, for NLP applications 128
    representing, numerically with vectors 131
large BERT model
    versus small BERT model 241-249
    versus TF-IDF evaluation 241-249
large language models
        (LLMs) 21, 77, 216, 230, 267, 277
    overview 196, 197
lemmatization 83, 148
lexicon 38
libraries
    exploring 53
    Keras 57, 58
    NLP libraries 58
    NLTK 53
    spaCy 55
Librispeech
    URL 75
linear scale 122
Linguistic Data Consortium
    reference link 30
    URL 75
log scale 122

long short-term memory (LSTM) 192, 197
loss parameter 187
low-resourced languages 5

M

machine learning (ML) 51, 179
machine translation 13
maintenance costs
    considering, for natural language
        applications 31, 32
Markdown 173
Matplotlib
    URL 59, 103
    used, for visualizing data 103-105
message boards 71
metadata 73
    annotation 73
metric for evaluation for translation with
        explicit ordering (METEOR) 232
metrics parameter 188
model.compile() method
    parameters 187
model.summary() method 188
multilayer perceptrons (MLPs) 179
    for classification 183-190
multimodal applications 23
multimodal interaction 285
Mycroft 8

N

Naïve Bayes 163
    used, for classifying documents 165
    used, for classifying texts 165, 166
named entity recognition
        (NER) 41, 56, 82, 191, 200

natural language 3
    basics 4, 5
natural language applications
    development costs, considering 31
    maintenance costs, considering 31, 32
natural language generation (NLG) 6, 20
natural language processing
        (NLP) 3, 6, 69, 96, 143, 163, 215
    applications, avoiding 26
natural language processing
        (NLP) application
    approaches, fitting to task 126
    approaches, selecting 126
    interactive applications 126
    non-interactive applications 126
natural language processing
        (NLP) approaches
    computational efficiency, considering 127
    data, initializing 126
    data requirement 126
    initial studies 128
    syntax 127
    vocabulary 127
natural language toolkit (NLTK) 39, 74, 266
    installing 55
    URL 53
    using 53-55
natural language understanding
        (NLU) 81, 95, 179, 196, 275, 276
    benefit 21
    problems, identifying 18-21
    versus graphical user interface 28
negative movie reviews
    versus positive movie reviews 106-109
neural networks (NNs) 58, 179-183
    activation function 181
    backpropagation 181

batch 181

connection 181

convergence 181

dropout 181

early stopping 181

epoch 181

error 182

exploding gradients 182

FCNN 182

forward propagation 182

gradient descent 182

hidden layer 182

hyperparameters 182

input layer 182

layer 182

learning 182

learning rate/adaptive learning rate 182

loss 182

MLP 182

neuron (unit) 182

optimization 182

output layer 182

overfitting 182

underfitting 183

vanishing gradients 183

weights 183

neurons 44, 180

new data generation

from crowdworkers 269

from LLMs 267-269

from rules 266, 267

ngram 110

NLP applications, language representation

symbolic representations 128-130

NLP libraries

one sentence, processing 62

selecting 58

useful packages 59

NLP-progress

URL 233

NLU application 20

dealing, with counterfactuals 22, 23

dealing, with hypotheticals 22, 23

dealing, with possibilities 22, 23

deciding, flowchart 33, 34

for identifying, user's need 25

integrating, broad general/
expert knowledge 23-25

multimodal 23

multiple languages, identifying 25, 26

system, to use judgment/common sense 22

viewing 21

NLU model

deploying 273

NLU performance problems

occurring, post deployment 273, 274

NLU system

evaluating 229-231

NLU technologies

selection considerations 46, 47

NLU technology and research, future 283

bias, detecting and correcting 285

multimodal interaction 285

NLU technologies, extending
to new languages 283

real-time speech-to-speech translation 284

non-interactive applications 7, 10, 14

analytics 12

authorship 13

classification 11

document retrieval 12

fake news detection 12

grammar correction 13

information extraction 13

sentiment analysis 11

spam detection 11, 12

summarization 13
translation 13
NumPy
URL 59

O

one-hot encoding 137
ontologies 149-151
usage 149
OpenAI
URL 212
optimizer parameter 188
oversampling 91, 126, 265

P

pandas
URL 59, 103
used, for visualizing data 103-105
parsing 39, 130, 152
partially annotated data
making, with weak supervision 226, 227
part of speech (POS)
tagging 38, 39, 54, 82, 84, 91
personally identifiable information (PII) 75
pipeline 42, 91
positive movie reviews
versus negative movie reviews 106-109
pragmatic analysis 41
precision 236
preprocessing techniques
advantages and disadvantages 91, 93
selecting 91
pre-trained models 45, 198
pretraining 198

Python
for NLP 15
installing 50, 51
URL 50
PyTorch
URL 58

R

RASA
URL 9
recall 236
recall-oriented understudy for gisting evaluation (ROUGE) 232
receiver operating characteristic (ROC) 237
rectified linear unit (ReLU) 181
recurrent neural networks (RNNs) 44, 179, 191
example 192
unit 192
regular expressions (regexes) 55, 145
strings, parsing 145, 147
strings, recognizing 145, 147
strings, replacing 145, 147
tips, for using 148
used, for analyzing text 26
ReLU function
benefits 186
representation 163
robust measurement 120
rule-based approaches 38
grammar 39
lexicons 38
parsing 39
part-of-speech tagging 38
pipelines 42

pragmatic analysis 41
semantic analysis 40
words 38
rule-based techniques 143
rules 143
benefits 144
left-hand side (LHS) 152
right-hand side (RHS) 152
runtime data
privacy, ensuring 76

S

scikit-learn
URL 58, 59
Seaborn
URL 59, 103
used, for visualizing data 103-105
semantic analysis 40, 155, 156
semantically similar documents, grouping
data representation 217
semantics 151
Sentence Bert (SBERT) 220, 270
sentence-level analysis 151
semantic analysis 155, 156
slot filling 156-159
syntactic analysis 152
Sentence Polarity Dataset 96, 97
sentences 151
sentiment analysis (SA) 11, 56
shared tasks 233
shared tasks paradigm
benefits 233
sigmoid function
using 187
slot filling 156-159
with CRFs 171, 172
slot labeling 191

slot-tagged data
representing 172-177
small BERT model
versus large BERT model 241-249
versus TF-IDF evaluation 241-249
spaCy
using 55-57
spaCy id attribute
using 159, 160
spam detection 11, 12
speech data
inter-annotator agreement 74
transcription 74
Speechmatics
reference link 284
speech recognition 6
speech-to-speech translation 284
reference link 284
speech-to-text 6
spoken language 4
Statistica
URL 127
statistical methods 172
statistical significance 240
of differences 240, 241
stemming 83
stopwords 65, 86, 101, 221
parameters 221, 222
strings
parsing, with regular expressions 145, 147
recognizing, with regular
expressions 145, 147
replacing, with regular expressions 145, 147
subordinate 149
superordinate 149
supervised learning 216
support vector machines (SVM) 44, 163
used, for classifying documents 169-171

symbolic representations **128-131**
synsets **150**
syntactic analysis **152**
   context-free grammars (CFGs) 152-155
syntax **151**
system issues
   figuring out 254
system issues, in initial development
   category balance, checking 254-257
   checking, for weak classes 259-263
   initial evaluations, performing 257, 258

T

TensorFlow
   URL 57
TensorFlow Hub
   reference link 200
term frequency-inverse document frequency
   (TF-IDF) **133, 217, 191, 241, 281**
   used, for representing documents 165
term frequency (TF) **133**
text
   analyzing, with regular expressions 26
   classifying, with Naïve Bayes 166
text classification methods
   comparing 241
   large BERT model 248, 249
   small BERT model 241-245
   small BERT model, comparing 244
   TF-IDF evaluation 246, 247
text document dataset **96, 97**
text preprocessing pipelines
   using 91
Text Retrieval Conference (TREC) **254**
text-to-speech **6**
TF-IDF/Bayes classification
   example 166-168

TF-IDF evaluation
   versus large BERT model 241-249
   versus small BERT model 241-249
topic modeling
   with clustering techniques 217
   with label derivation 217
traditional machine-learning approaches **42**
   classification 44
   documents, representing 43
   drawbacks 42
training data **29-31**
   privacy, ensuring 76
training process **180**
   epochs 180
transformers **197**
   attention, applying 198
   overview 196, 197
translation **9, 13**
trigram **110**
tuning
   modifying, to improve
      performance 190, 191

U

undersampling **91, 126, 265**
Uniform Manifold Approximation and
   Projection (UMAP) **218, 222**
   parameters 222
unsupervised learning **216**
user testing **239, 240**

V

validation data **233**
vectors
   used, for representing language
      numerically 131

vectors, for document representation 131
    binary bag of words 131, 132
    count bag of words 132, 133
    term frequency-inverse document
        frequency (TF-IDF) 133-137
visualization 218, 224, 225
    developing, considerations 119-123
    information, using to make decisions
        about processing 123

W

weak supervision
    tactics 227
    used, for making partially
        annotated data 226, 227
wildcard characters 27
Wizard of Oz (WoZ) method 72
Word2Vec 137-141, 217
word cloud 64, 105, 106
    URL 59

word embeddings 137
word error rate 235
word frequency distributions 98-103
word-level analysis 148
    lemmatization 148
    ontologies 149-151
WordNet 149
words
    representing, with
        context-dependent vectors 141
    representing, with
        context-independent vectors 137
World Wide Web (WWW) 23
written language 4

Z

zero-shot learning 210
Zipf's law 105

www.packtpub.com

Subscribe to our online digital library for full access to over 7,000 books and videos, as well as industry leading tools to help you plan your personal development and advance your career. For more information, please visit our website.

Why subscribe?

- Spend less time learning and more time coding with practical eBooks and Videos from over 4,000 industry professionals

- Improve your learning with Skill Plans built especially for you

- Get a free eBook or video every month

- Fully searchable for easy access to vital information

- Copy and paste, print, and bookmark content

Did you know that Packt offers eBook versions of every book published, with PDF and ePub files available? You can upgrade to the eBook version at www.packtpub.com and as a print book customer, you are entitled to a discount on the eBook copy. Get in touch with us at customercare@ packtpub.com for more details.

At www.packtpub.com, you can also read a collection of free technical articles, sign up for a range of free newsletters, and receive exclusive discounts and offers on Packt books and eBooks.

Other Books You May Enjoy

If you enjoyed this book, you may be interested in these other books by Packt:

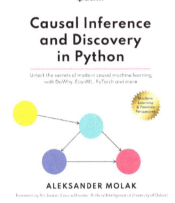

Causal Inference and Discovery in Python

Aleksander Molak

ISBN: 9781804612989

- Master the fundamental concepts of causal inference
- Decipher the mysteries of structural causal models
- Unleash the power of the 4-step causal inference process in Python
- Explore advanced uplift modeling techniques
- Unlock the secrets of modern causal discovery using Python
- Use causal inference for social impact and community benefit

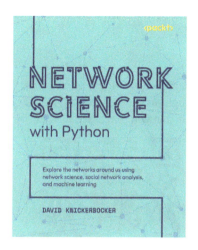

Network Science with Python

David Knickerbocker

ISBN: 9781801073691

- Explore NLP, network science, and social network analysis
- Apply the tech stack used for NLP, network science, and analysis
- Extract insights from NLP and network data
- Generate personalized NLP and network projects
- Authenticate and scrape tweets, connections, the web, and data streams
- Discover the use of network data in machine learning projects

Packt is searching for authors like you

If you're interested in becoming an author for Packt, please visit `authors.packtpub.com` and apply today. We have worked with thousands of developers and tech professionals, just like you, to help them share their insight with the global tech community. You can make a general application, apply for a specific hot topic that we are recruiting an author for, or submit your own idea.

Share your thoughts

Now you've finished *Natural Language Understanding with Python*, we'd love to hear your thoughts! Scan the QR code below to go straight to the Amazon review page for this book and share your feedback or leave a review on the site that you purchased it from.

https://packt.link/r/1-804-61342-8

Your review is important to us and the tech community and will help us make sure we're delivering excellent quality content.

Download a free PDF copy of this book

Thanks for purchasing this book!

Do you like to read on the go but are unable to carry your print books everywhere?

Is your eBook purchase not compatible with the device of your choice?

Don't worry, now with every Packt book you get a DRM-free PDF version of that book at no cost.

Read anywhere, any place, on any device. Search, copy, and paste code from your favorite technical books directly into your application.

The perks don't stop there, you can get exclusive access to discounts, newsletters, and great free content in your inbox daily

Follow these simple steps to get the benefits:

1. Scan the QR code or visit the link below

https://packt.link/free-ebook/9781804613429

2. Submit your proof of purchase

3. That's it! We'll send your free PDF and other benefits to your email directly

www.ingramcontent.com/pod-product-compliance
Lightning Source LLC
Chambersburg PA
CBHW062104050326
40690CB00016B/3197